FOOD ISSUES, POLICIES, AND SAFETY CONSIDERATIONS

FOCUS ON AFLATOXINS RESEARCH

FOOD ISSUES, POLICIES, AND SAFETY CONSIDERATIONS

Additional books in this series can be found on Nova's website under the Series tab.

Additional e-books in this series can be found on Nova's website under the eBooks tab.

FOOD ISSUES, POLICIES, AND SAFETY CONSIDERATIONS

FOCUS ON AFLATOXINS RESEARCH

PEDER JAN
AND
STEN SYLVESTER
EDITORS

nova
science publishers
New York

Library of Congress Cataloging-in-Publication Data

ISBN: 978-1-53612-569-6

Published by Nova Science Publishers, Inc. † New York

CONTENTS

PREFACE

In the first chapter, Yin-Hui Leong, Nurul Izzah Ahmad, and Rahmat Awang, provide an overview of literature on recent progress in aflatoxins research including health effects and human exposure through alimentary or dermal pathways. Next, Mufida Aljicevic MD, PhD, and Damir Ljuboja present a study with the goal to decide whether foods of plant origin have quantifiable concentrations of aflatoxin B1. Additionally, they aimed to examine whether the urine of healthy participants contains measurable aflatoxin B1 in the second chapter. The third chapter by Hsin-Bai Yin, Anup Kollanoor-Johny, and Kumar Venkitanarayanan pinpoints the potential of various natural approaches and their applications in controlling AF contamination at pre- and post-harvest stages of agriculture. During the fourth and final chapter, Edlayne Gonçalez, PhD, Janaina Lara da Silva, MD, Tatiana Emy de Freitas, MD, Christiane Ceriani Aparecido, PhD, Daiane Hansen, PhD, and Joana D'arc Felicio, PhD examine the presence of toxigenic species in samples for lactating goats, determining whether this indicates a potential risk of contamination of dairy products.

Chapter 1 – Aflatoxins are the most important mycotoxins produced predominantly by specific strains within species of the Aspergillus genus. Aflatoxins contaminate a vast variety of foods and feeds: from agriculture to the plate of consumers. It is a serious problem worldwide, affecting the economy and health of nations. Moreover, aflatoxin producing fungi, particularly *A.* flavus, is expected to elevate occurrences of aflatoxins contamination with global warming and climatic change in areas where they were not previously prevalent. This review summarizes literature data on recent research progress on aflatoxins, including its effects on health, occurrence, and human exposure via alimentary and inhalation or dermal pathways. In addition, it provides an overview of metabolism, biomarker-based methodologies and internal body burden to characterize total human exposure. It also discusses current toxicologically-based values and different approaches used for risk assessment.

Chapter 2 – Molds are microscopic fungi that produce metabolites and release them to the surface where growth occurs. Mycotoxins are such metabolites, with over 400 different toxins produced by approximately 350 species of fungi. Aflatoxins are mycotoxins produced by *Aspergillus flavus* under favorable temperature and humidity conditions. Aflatoxins belong to a group of extremely toxic mycotoxins, and are therefore categorized in as Group I carcinogens by the International Agency for Research on Cancer (IARC). There are four main types of aflatoxins: B1, B2, G1 and G2, as well as two additional metabolic products: M1 and M2. The IARC has placed aflatoxin B1 on the list of human carcinogens. Man is exposed to the effects of aflatoxins via consumption of food that is contaminated by products of developed fungi.

Aflatoxin B1 is the strongest natural carcinogen for the liver and aflatoxin B1 and hepatitis B (HBV) are considered as major risk factors in the etiology of hepatocellular carcinoma (HCC).

The aim of this work was to determine if foods of plant origin present in our market have quantifiable concentrations of aflatoxin B1. A secondary objective was to examine whether the urine of healthy participants contains measurable aflatoxin B1, and compare this control to a population of patients with HBV and HCC.

Laboratory testing can detect the presence of aflatoxins in samples of food, as well as in the urine of patients. In the authors of aflatoxin B1, samples were divided into three groups for analysis. The first group consisted of 32 samples foods of plant origin that were available in the marketplace, the second experimental group consisted of the urine of 30 patients with HCC and HBV, and the third (control) group consisted of the urine 30 "healthy" patients. All samples were processed using the Aflatoxin B1 Flow Through Rapid Test kit by Euro-Diagnostica. All samples were tested by Sensitive Aflatoxin B1 enzyme immunoassay (EIA) for screening and quantitative analysis. From a total of 32 food samples tested, first screening analysis with Flow Through Rapid Test yielded only 4 positive results, while EIA detected 11 positive results with quantitative values of aflatoxin B1 ranging from 0.23 to 11.67 ppb. In the healthy control patients and the experimental group, the values of aflatoxin B1 in the urine samples were below the detectable level of screening test (2 ppb) via the Flow Through Rapid Test. However, with Sensitive EIA testing we obtained values of 0.05 to 0.15 ppb of aflatoxin B1 in the control urine samples, and 0.05 to 0.26 ppb in the experimental group. Comparing the results clearly shows that the Aflatoxin B1 Sensitive EIA is more sensitive than Aflatoxin B1 Flow Through Rapid Test. The Rapid test is a simple and quick screening method, but is not sufficiently sensitive for quantifying aflatoxin B1 levels.

Of note, all of tested urine samples of both groups (experimental and control) showed the presence of aflatoxins B1. However, the concentration of aflatoxins was lower in the control group (maximum value of 0.15 ppb), relative to the experimental group (maximum value of 0.26 ppb).

These results suggest that aflatoxin-contaminated foods readily available in the marketplace serve as a vector for the introduction of small quantities of aflatoxins into hosts. These aflatoxins may pose a cumulative threat to human health over time given their greater concentration in patients with HBV and HCC. Thus, further research into aflatoxins and their role in the pathophysiology of HCC is indicated.

Chapter 3 – Aflatoxins (AF) are fungal toxic metabolites produced by *Aspergillus flavus* and *Aspergillus parasiticus* that frequently contaminate a variety of crops, including peanut, corns, and cottonseed. Aflatoxin contamination causes an estimated average economic loss of $1 billion annually to the US agriculture. Additionally, contamination of feed ingredients with AF is a major concern to animal agriculture due to their deleterious effects on livestock such as reduced animal performance and increased mortality. Moreover, the consumption of AF-contaminated animal products negatively affects public health due to their carcinogenic and hepatotoxic effects. The common practices for detoxification of AF in crops include prolonged heating under pressure and the use of chemicals such as oxidizing agents and ammonia. However, nutrient losses associated with heat processing and public health safety concerns on chemical residues have limited the applicability of these methods. Moreover, the inclusion of AF-binding adsorbents in feed to protect livestock from the harmful effects of AF has been shown to impair nutrient utilization and mineral absorption in livestock. Thus, it is critical to develop effective strategies for controlling AF in crops, feed ingredients and livestock. Extensive studies have identified several natural plant- and animal-derived compounds exhibiting potent antifungal or antitoxigenic properties against toxin producing molds. This chapter highlights the potential of various natural approaches, including atoxigenic or probiotic microorganisms, phytochemicals, and organic compounds in reducing AF as well as the mechanisms behind their antitoxigenic effect. In addition, their potential application in controlling AF contamination at pre- and post-harvest stages of agriculture is discussed.

Chapter 4 – Aflatoxins are highly toxic and carcinogenic metabolites produced by *Aspergillus* species in food, feed and agricultural commodities. Several studies on aflatoxins in feed and bovine milk have been reported, but in feed and milk the goats are scarce. This study aimed to evaluate the presence of aflatoxins in diets intended for dairy goats and their residues in milk and identify the mycobiota present in the feed and forage. *Aspergillus flavus* and *A. parasiticus* strains were isolated and characterize based on a morphological and molecular characterization and mycotoxigenic ability. The goat dairy diets were collected monthly from 10 goat milk farms, totaling 128, 106 and 120 samples of concentrate, forage and milk, respectively. For the isolation of the mycobiota the surface plating method was used. *Aspergillus*, *Penicillium*, and *Fusarium* were the main fungi producing mycotoxins isolated. The morphological and molecular characterization and mycotoxigenic ability were used for *A. flavus* and *A. parasiticus* identification. The *Aspergillus* spp. from feed

39% produced aflatoxins B_1 and B_2, and 2.0% of the strains produced aflatoxins B_1, B_2, G_1, and G_2. The strains from forage were producers of aflatoxins B_1 and B_2 (37%), whereas 49% have shown no toxigenic ability. The *aflD* and *aflR* genes were used by PCR and PCR-RFLP, respectively. Aflatoxins have been extracted and purified in an immunoaffinity column specific for aflatoxins B_1, B_2, G_1, G_2 and M_1. The analyses for detection and quantification of toxins were performed by high-performance liquid chromatography. From the analyses of the feeds, 2.34% of these were contaminated with aflatoxins, but aflatoxin M_1 was not detected in the milk samples. Despite low contamination of feed, the presence of toxigenic species in samples for lactating goats indicates a potential risk of contamination of dairy products, if they are exposed to environmental conditions favorable to fungal growth and mycotoxin production.

In: Focus on Aflatoxins Research
Editors: P. Jan and S. Sylvester
ISBN: 978-1-53612-569-6

Chapter 1

OCCURRENCE, HUMAN EXPOSURE AND THE CURRENT TRENDS OF EXPOSURE MEASUREMENTS FOR AFLATOXINS

Yin-Hui Leong[1,*], Nurul Izzah Ahmad[2] and Rahmat Awang[1]
[1]National Poison Centre of Malaysia,
Universiti Sains Malaysia, Minden,
Penang, Malaysia
[2]Environmental Health Research Center,
Institute for Medical Research, Jalan Pahang,
Kuala Lumpur, Malaysia

ABSTRACT

Aflatoxins are the most important mycotoxins produced predominantly by specific strains within species of the Aspergillus genus. Aflatoxins contaminate a vast variety of foods and feeds: from agriculture

[*] Corresponding author. Tel: +604 653 2077; Fax: +604 656 8417,
Email: yinhui_leong@yahoo.com.

to the plate of consumers. It is a serious problem worldwide, affecting the economy and health of nations. Moreover, aflatoxin producing fungi, particularly *A.* flavus, is expected to elevate occurrences of aflatoxins contamination with global warming and climatic change in areas where they were not previously prevalent. This review summarizes literature data on recent research progress on aflatoxins, including its effects on health, occurrence, and human exposure via alimentary and inhalation or dermal pathways. In addition, it provides an overview of metabolism, biomarker-based methodologies and internal body burden to characterize total human exposure. It also discusses current toxicologically-based values and different approaches used for risk assessment.

INTRODUCTION

Aflatoxins are a group of mycotoxins produced as highly toxic secondary metabolites via a polyketide pathway by various species of Aspergillus section Flavi. Over 400 mycotoxins are known and the major classes of mycotoxins are aflatoxins, trichothecences, fumonisins, zearalenone and ochratoxin A. The frequent occurrence of these food-borne toxins and their severe effects on animal and human health has been recognized (Hussein and Brasel, 2001). Aflatoxins are probably the most extensively studied mycotoxins. The name aflatoxin is an abbreviation of "a" for *Aspergillus* genus and "fla" for the species *flavus*, and the word "toxin" (meaning poison) (Ellis et al., 1991). Currently, more than 14 different types of aflatoxins have been identified. Naturally occurring aflatoxins are: aflatoxin type B1, B2, G1 and G2 (noted AFB1, AFB2, AFG1 and AFG2 respectively).

AFB1 and AFB2 are produced typically by a fungi belonging to *A. flavus* while AFG1 and AFG2 are produced by *A. parasiticus*. However, *A. bombycis*, *A. ochraceoroseus*, *A. nomius*, and *A. pseudotamari* are also reported to be aflatoxin-producing species although their occurrence is less common (Klich et al., 2000; Kurtzman et al., 1987; Peterson et al., 2001). *A. flavus* occurs throughout the world, whilst *A. parasiticus* is restricted mainly to Africa, South America, Central America, and North America (Turner, 2013). Aflatoxins were discovered in the 1960s when a series of outbreaks in poultry and fish occurred in different parts of the world. One of the famous outbreaks was the sudden death of more than 100,000 turkeys and other farm animals (turkey "X" disease) in UK where the cause was attributed to *A. flavus* contaminated peanut meal (Bennett and Klich, 2003). Aflatoxins occur mostly in tropical regions where hot and humid climates promote fungal growth. Poor harvesting practices, improper storage, and less than optimal conditions during transportation and marketing can also contribute to proliferation of fungal and increasing the risk of aflatoxins production (Bhat and Vasanthi, 2003). In fact, this contamination can occur in any stage of food production, from pre-harvest to storage (Turner et al., 2005).

Structure and Toxicity

Aflatoxins display fluorescence under ultraviolet light and are grouped into AFB1 or AFB2 for blue, and AFG1 or AFG2, for green, where number 1 or 2 refer to the chromatographic mobility during thin-layer chromatography (Bennett and Klich, 2003). Aflatoxins are a group of structurally related difuranocoumarins and attached to either pentanone (AFB1 and AFB2) or a six-membered lactone (AFG1 and AFG2) (Figure 1). Aflatoxin M1 and M2 (AFM1, AFM2), are the metabolites of AFB1 and AFB2 respectively, and are found in both the meat and milk of animals whose feedstuffs have been contaminated by AFB1 and AFB2 (de Ruyck et al., 2015; Iqbal et al., 2015). From the four major aflatoxins, AFB1 is the most toxic and most prevalent compound, followed by AFG1, AFB2 and AFG2 with decreasing toxicity. The carcinogenic potency of AFM1 is around 10-fold lower than that of AFB1.

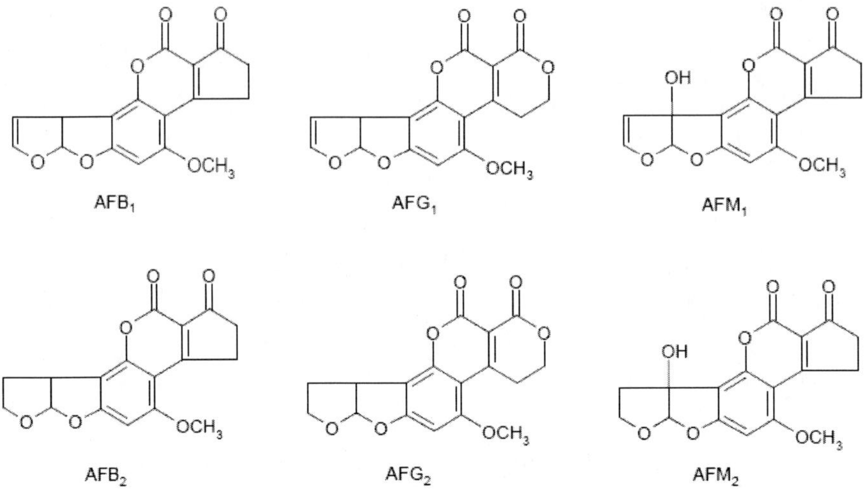

Figure 1. Chemical structures of major aflatoxins (AFB1, AFB2, AFG1, AFG2, AFM1 and AFM2).

HEALTH EFFECTS

Diseases caused by aflatoxin consumption are collectively referred to as aflatoxicosis. Acute aflatoxicosis results in death while chronic aflatoxicosis causes liver cancer, immune suppression, and other pathological conditions (Hsieh, 1988; Sherif et al., 2009). Symptoms of acute aflatoxicosis may include high fever, vomiting, hemorrhage, rapid progressive jaundice and swollen livers (Fung and Clark, 2004). The International Agency for Research on Cancer (IARC) has classified aflatoxins (including AFB1, B2, G1, G2 and M1) as Group I human carcinogen (IARC, 2012). It has been shown to produce DNA-adducts, gene mutation and chromosomal anomalies in humans and animals *in vivo* and *in vitro* (IARC, 1993). Their role in hepatocarcinogenesis often in conjunction with hepatitis B is well established (Sun et al., 2013; Wild and Hall, 1998; Wild and Turner, 2002). Additionally, high prevalence of infectious diseases, such as diarrhea (Mohammad-Hasani et al., 2016), and infection with HIV or progression to AIDS due to immunosuppressive effects of aflatoxin exposure have been published (Natamba et al., 2016). There is also evidence of associations with Reye's syndrome (encephalopathy with severe lesions in kidney and liver following influenza or varicella), Kwashiorkor (severe protein-energy malnutrition disease), and acute hepatitis (Wild and Hall, 1998). An illness known as "cirrhosis of Indian childhood" is also partly due to an aflatoxin poisoning. Aflatoxins can cross the human placenta and exposure to it in early life is associated with impaired growth, particularly stunting (Gong et al., 2002; IARC, 2002; Turner, 2013; Wu and Khlangwiset, 2010). Recent studies have suggested a strong link between childhood stunting, a developmental disorder with various short-term and long-term adverse health outcomes, and chronic dietary exposure to mycotoxins, particularly AFB1 (de Onis and Branca, 2016; Groopman et al., 2014; Watson et al., 2015).

In studies on acute outbreaks of aflatoxicosis in Kenya (Azziz-Baumgartner et al., 2005; Ngindu et al., 1982), India (Krishnamachari et al., 1975) and Thailand (Bourgeois et al., 1971), staple foods contaminated with 5,000 ppb or above of aflatoxins were linked to fatality while daily consumption of foods with greater than 1,000 ppb was linked to aflatoxicosis. Based on the observed levels of contamination in the studies, the intake of total aflatoxins leading to fatality risk can be estimated to be greater than 1 mg/kg, or in excess of 20 μg/kg bw/day in adults (Wild and Gong, 2010). According to Turner et al., (2012a, b), populations that are particularly likely to have chronic and high levels of exposure are typically the poor, those with limited dietary variety and are reliant on maize or groundnuts as dietary staples. In Malaysia, acute outbreaks of aflatoxicosis were reported (Chao et al., 1991; Lye et al., 1995). The severity of the illness, however, differed from previously reported outbreaks described in Kenya and India. The offending food was a Chinese noodle called "Loh See Fun." The poisoning, in 1988, during a religious festival claimed the lives of 13 children in the state of Perak, Malaysia. It was alleged that up to 3 mg of aflatoxin could have been present in the contaminated noodles, based on estimates from the Kenyan experience (Lye et al., 1995).

Among farm animals, ingestion of aflatoxins contaminated feeds can lead to substantial loss in productivity and meat quality (Bonomi et al., 1994). Symptoms of acute aflatoxicosis in mammals include lethargy, ataxia, rough hair coat, and enlarged fatty liver (Nibbelink, 1986). For chronic exposures, the major symptoms are reduced intake and milk production, jaundice and decreased appetite (Nibbelink, 1986). Reduced growth rate is possibly the most obvious indication for chronic aflatoxicosis (Pier, 1992) and is related to disturbances in protein, carbohydrate and lipid metabolism (Cheeke and Shull, 1985).

Occurrence of Aflatoxins and Human Exposure

Contamination by aflatoxins is regulated worldwide to curb exposure to animals and humans (Wu and Guclu, 2012). Important mycotoxins regulated in foods and feeds legislated by the EU has been reviewed by Fromme et al. (2016). A review on monitoring studies of occurrences of aflatoxins in food products has demonstrated that aflatoxins are frequently present in food products at levels that are of significant concern for consumer protection (Stroka and Anklam, 2002). According to an EU screening from 2000 – 2006, aflatoxins were detected in 26% of all analysed foods (EFSA, 2007). Aflatoxins were commonly found to have contaminated a wide range of key agricultural commodities such as, cereals (maize, sorghum, rice, wheat), spices (chili, turmeric, black pepper), oilseeds (peanut, cottonseed), tree nuts (almond, pistachio, walnut), dried fruits and milk (human and animals) (Bhat et al., 2010; Iqbal et al., 2015, Masood et al., 2015). High concentration of AFM1 in milk, between 28 and 1012 ng/kg was reported in some European Union countries (Markaki and Melissari, 1997; Martin and Martin, 2000). In Serbia, AFM1 was detected in milk in concentrations above the maximum residue levels of 0.05 mg/kg defined by the European Union (EU) Regulation (Kos et al., 2014) and in infant follow-on milks as well as infant formulas, ranging from 55 to 201 ng/kg in Turkey (Er et al., 2014). All powdered milk samples collected in Argentina and Brazil had AFM1 ranging from 100 - 920 ng/kg (Londono et al., 2013). Two years ago Cano-Sancho et al. (2015) reported that 33% of examined yoghurt samples in Spain exceeded the EU limit.

Occurrence of aflatoxins in several developed countries e.g., USA, Sweden, UK and others was reported in 1990s (Moss, 2002) but there are evidence of more incidents in the developing countries. Occurrences of AFB1 are expected to become even more prevalent with climate change in Europe and in countries with temperate climate that did not previously have this problem (Battilani et al., 2016; Paterson and Lima, 2010; Tirado et al., 2010). In developing countries, problems of aflatoxins might be largely due to less sophisticated food handling or preservation technology including unsatisfactory control or regulation over food quality and safety compared to more developed countries. In addition to this, the fungi producing aflatoxins are prevalent in areas of high humidity and temperature therefore crops in tropical and subtropical regions are more susceptible to contamination. Several factors influenced fungi growth such as, availability of nutrients, temperature, acidity/pH, moisture, water activity, and other associated factors (WHO, 1999). Table 1 shows a list of notification of aflatoxins in foods recently reported by the Rapid Alert System for Food and Feed (RASFF) (https://webgate.ec.europa.eu/rasff-window/portal/). The European Food Safety Authority (EFSA) operates RASFF as a tool for information exchange on food and feed consignments when foods at risks to human health are identified, recalled, seized, rejected or withheld.

Occurrences of aflatoxins in developing countries have been documented. A study by Roy et al. (2013) shows the presence of AFB1 in pooled samples of eight commonly consumed food commodities and in poultry feed in Bangladesh, 5 out of 8 food commodities among them exceeding the EU regulatory limits for aflatoxins.

Table 1. Detection of aflatoxins in foods by RASFF from March to mid-June 2017

Food category	No. of cases	Origin country	Range (μg/kg)
Cereal and bakery product	2	Italy	AFB1: 3 – 54
	1	Slovenia	Total: 5.6 – 67
	1	Denmark	
	5	Switzerland	
	1	France	
	1	Poland	
Fruit and vegetables	2	Germany	AFB1: 10.1 – 54.6
	1	Spain	Total: 12.8 – 102.4
	2	France	
Herbs and spices	1	Ireland	AFB1: 6.1 – 48.1
	2	Netherlands	Total: 10.8 – 107.1
	1	Finland	
	1	Denmark	
	1	Greece	
	1	Sweden	
Nuts, nut products and seeds	12	Netherlands	AFB1: 2.4 – 3125
	7	Belgium	Total: 4.0 – 3232
	23	Germany	
	9	France	
	1	Luxembourg	
	5	Bulgaria	
	12	Italy	
	3	Portugal	
	5	Poland	
	14	Spain	
	6	United Kingdom	
	3	Greece	
	2	Cyprus	
	1	Sweden	
	1	Finland	
	1	Romania	
	2	Switzerland	

Search criteria: notified from 01/03/2017; notified until 15/06/2017; product type food; hazard category mycotoxins. (RASFF, 2017).

Incidence and level of aflatoxins B1, B2, G1 and G2 contamination in starch-based foods, commercial traditional herbal medicines, peanuts and corn products, rice and its by-products and baby foods from Malaysia, Indonesia, Philippines and Thailand have been reported (Abdullah et al., 1998; Ali et al., 1999; Ali et al., 2005; Noviandi et al., 2001; Razzazi-Fazeli et al., 2004; Sales and Yoshizawa, 2005; Tassaneeyakul et al., 2004). Fifty-seven percent of nuts and nut products from Penang, Malaysia, were also found to be aflatoxin-contaminated, in the range of 0.40 to 222 μg/kg (Leong et al., 2011). Sulaiman et al. (2007) reported that a very high concentration of total aflatoxins (0.85 – 762.05 μg/kg) was found in 145 shelled raw peanut samples collected in Perak state, Malaysia. AFM1 was found in 55.56% of 168 samples of raw milk with a mean concentration of 21.31 ng/l and 30% of them were higher than EU limit (>50 ng/L) in Southern Iran (Hashemi, 2016). However, the occurrence of AFM1 in milk and dairy products in Malaysia is still relatively low, only (7.5%) of the tested samples had the contamination level greater than the EU limit (Farah Nadira et al., 2017).

The primary route of aflatoxins exposure is oral consumption of aflatoxins contaminated foods, a prime food safety issue and public health concern particularly in developing countries. Exposure measurement of aflatoxins were extensively studied during the 1980s. According to Williams et al. (2004), an estimated 4.5 billion of the world's population are exposed to aflatoxins. It is estimated that human consumption of aflatoxins ranges between 0 and 30,000 ng/kg/day with an average intake of 10 to 200 ng/kg/day (Revankar, 2003). In Sub-Saharan Africa and South-East Asia, detection of aflatoxins is regularly detected. High exposures have been reported in East Africa, China and parts of South-East Asia (Wild and Gong, 2010). Egal et al. (2005) reported that 90% of young children in Benin and Togo, West Africa, are exposed to aflatoxins from consumption of contaminated maize and groundnuts. Daily human exposure to aflatoxin vary in different countries: an estimate of 4 – 184 ng/kg bw in several African regions, 12 – 2027 ng/kg bw in southern China, 7 – 53 ng/kg bw in Thailand, and comparably low of < 3 ng/kg bw in USA (Williams et al., 2004). In Malaysia, (Leong et al., 2011) reported an estimated daily dietary exposure to AFB1 in nuts and nut products as 0.36 ng/kg bw and 8.89 ng/kg bw, a representation of low and high-level of exposure. Human exposure to aflatoxins can result directly from ingestion of contaminated plant materials or products or indirectly from consumption of foods from animals previously exposed to aflatoxins in feeds. Aflatoxins resistance to processing and their stability during cooking also contribute to the frequency of dietary exposure (Miller, 1995). Although exposure to aflatoxins is mainly caused via alimentary route, toxicity occurring through inhalation or dermal pathways has been reported (Boonen et al. 2012; Degen 2011; Fromme et al., 2016; Kelman et al. 2004). Fromme et al. (2016) reviewed various investigations measuring the concentrations of mycotoxins (primarily aflatoxins and ochratoxin A) in the air or dust of agricultural environments, workplaces and in residences.

Aflatoxins are one of the most well documented mycotoxins in indoor environments (Kilburn, 2004; Zain, 2011). *A. flavus* and aflatoxins have been detected and identified in a cold-water storage tank by Paterson et al. (1997). A study by Hedayati et al. (2011) found that the fungal flora of tap water from an Iranian university hospital was a reservoir for fungi, in particular Aspergillus. *A. flavus* was also frequently identified and isolated in indoor air, especially in subtropical and tropical areas (Hedayati et al., 2010; Sepahvand et al., 2013). This species was recently discovered in vast amount in indoor air in Croatia and Hungary (Jakšić et al., 2014; Varga et al., 2013). However, most of studies have only identified the species but not the toxins produced as byproducts in the indoor environments. Table 2 shows several studies on indoor air identifying the *Aspergillus* species as among the predominant genera of fungi in selected locations. This suggests the potentially producing mycotoxins in indoor environments where the species were detected. According to Nielsen (2003), factors that affect mycotoxin production in buildings are, secondary metabolites and mycotoxins are species-specific, mixtures of metabolites can produce synergistic effects, certain mycotoxins can be masked as derivatives, and metabolite production is influenced by medium composition, temperature and water activity. These factors denotes that molds are likely to generate different metabolites when they grow on building materials and toxin production is affected by the fact that molds grow in mixed cultures with other fungi and bacteria in buildings (Nielsen, 2003).

Table 2. Some studies on indoor air environments of fungi contamination

Sampling locations	Fungi Species identified	Important findings	References
Airborne fungi collected in 6 countries (Croatia, Hungary, the Netherlands, Turkey Algeria & Thailand)	*Aspergillus spp (A. niger, A. welwitschiae, A. carbonarius, A. tubinggensis, A. luchuensis, A. piperis, A. uvarum, A. aculeatinus, A. neoniger, A. fijiensis, A. japonicas, A. trinidadensis)*	178 black *Aspergillus* isolates were identified. Species distribution differed among countries but in temperate climate is considerably similar. Most species diversity was observed in Thailand. Potentially ochratoxin-and fumonisin-producing fungi were detected in all countries	Varga et al. 2014
6 Cincinnati area single-family homes, south-West Ohio & Northern Kentucky, USA	*Penicilliun spp., Clodosporium spp., aspergillus spp., Yeast, Chrysosporium spp. Paecilomyces spp. & Sporothrix spp.*	The geometric means of indoor cultured fungal concentrations are 88 cfu/m3. 26 genera of cultured fungi were recovered from the indoor and outdoor samples.	Lee et al. 2006

Table 2. (Continued)

Sampling locations	Fungi Species identified	Important findings	References
		The indoor median viabilities of fungi is 55%. *Cladosporium* had the highest median value of cultivability (38%) followed by *Aspergillus/Penicillium* (9% and 2%) among predominant genera of fungi.	
Air-conditioned office buildings in Warsaw, Poland	*Filamentous fungi: Penicillium spp., Micrococcus spp., Penicillium chrysogenum Thom, Micrococcus roseus, P. commune Thom, Kocuria kristinae, P. citrinum Thom, S. epidermidis, P. marneffei, Thom, S. capitis, A. candidus Link, S. hominis, A. terreus Thom, S. sciuri, A. fumigates, S. warneri, Cladosporium spp., S. auricularis, Acremonium spp., S. saprophyticus, Mucor spp., Yeast: Rhodotorula mucilaginosa*	Median concentrations of bioaerosols in indoor and workplaces are 11 and 21cfu/m^3 respectively.	Szymczak & Gorny 2010
66 private households in the state of Styria, Austria	*Cladosporium spp., Penicillium spp., Alternaria spp., Aspergillus spp.,Auriobasidium spp. Schizophyllium spp., Wallemia spp., Eurotium spp., Mucor spp.*	Apartments visibly affected by mold, the median values are significantly higher than those of apartments without visible mold growth. The predominance *of Penicillium spp.* and *Aspergillus spp.*	Haas et al. 2007

Sampling locations	Fungi Species identified	Important findings	References
		($p < 0.001$) in indoor air. The total fungal concentration of *Penicillium* and *Aspergillus* in the air of apartments is recommended for assessing fungal exposure.	
Air-conditioned indoor environment (computer lab, hospitals, CT scan centres, etc. in Nizamabad, Andra Pradesh, India.	*Aspergillus spp. (A. niger, A. oryzae, A. fumigatus, A. terreus, A. nidulans, A. versicolor & A. parasiticus.) Penicillium citrinum, Fusarium oxysporum, Trichoderma viride, Neurospora crassa, Alternaria alternata.*	*A. niger* is present in 80% of the locations. Some of the fungi isolated in this study could be opportunistic fungal pathogens, such as *A. fumigatus, A. niger* and *Penicillium citrinum*, and were found to be allergenic.	Khan et al. 2009
10 different Hammams (traditional public baths) in Batna city, Northeast Algeria.	*Cladosporium, Fusarium, Rhizopus, Mucor and Alternaria, C. albicans, C. glabrata, C. tropicalis, C. lipolytica, Geotrichum sp., Trichosporon sp., Rhodotorula sp. and Cryptococcus sp.*	The most abundant molds are *Penicillium spp.* (45.12%) followed by *Aspergillus spp.* (28.80%). *C. albicans* is the common isolated yeast (35.14%). This study indicates that Hammams present a potential source of pathogenic fungi that could impose a serious threat to public health.	Benammar et al. 2017
Different cereal storage and food go-downs in Visakhapatnam, Andhra Pradesh, India.	*Alternaria alternate, A. solani, Aspergilus candidus, A. flavus, A. fumigatus, A. niger, A. parasiticus, A. terreus, A. versicolor, Botrytis spp., Cephalosporium spp., Cercospora spp.,*	Fungal spores identified in these areas are 1.7 x 103 to 4.2 x 106. The most dominating species are *Mucor, Rhizopus, Aspergillus, Penicillum, Alternaria, Cladosporium,*	Reddy et al. 2012

Table 2. (Continued)

Sampling locations	Fungi Species identified	Important findings	References
	Cladosporium cladosporioides, Colletotrichum spp., Curvularia affinis, Curvularia lunata, Fusarium moniliforme, F. oxysporium, F. solani, Helminthosporium spp., Mortierella zonata, Mucor microspores, Mucor racemorus, Penicillium spp., Rhizopus stolonifera, Stachybotrys spp., Trichoderma spp., Trichothecium spp., Verticillium spp., Yeast cells.	*Trichoderma, Fusarium, Pseudomonas, Proteus vulgari* and *Enterobacter aerogenes.* Strains of *Aspergillus* producing as high as 1927.3µg/kg aflatoxin.	

METABOLISM

Aflatoxin metabolism are extensively studied in animals and humans and several comprehensive reviews have been published (Eaton and Gallagher, 1994; Guengerich et al., 1998; Wild and Turner, 2002). Aflatoxin is metabolized by liver cytochrome P450 enzymes following ingestion to form highly reactive 8, 9-epoxide that subsequently reacts with DNA to generate guanine adducts (Bennett and Klich, 2003; Groopman et al., 1993; Walton et al., 2001), or with serum albumin to generate lysine adducts (Sabbioni et al., 1990). Aflatoxin can be hydroxylated to other metabolites such as AFM1 and AFQ1. The pathways and effects for aflatoxin in animal metabolism are shown in Figure 2. The differences in susceptibility to aflatoxin across species and between people depend largely on the fraction of the dose that is directed into the various possible pathways, with harmful "biological" exposure being the result of activation to the epoxide and the reaction of the epoxide with proteins and DNA. There is also evidence that the fractions that follow the different possible pathways are influenced by dosage, perhaps because of the saturation of the most chemically competitive processes (Eaton et al., 1993). Susceptibility to aflatoxin is greatest in the young, and there are very significant differences between species, individual of the same species (according to their differing abilities to detoxify aflatoxin by biochemical processes), and the sexes (according to the concentrations of testosterone). The toxicity of aflatoxin also varies according to many nutritional factors (Ayub and Sachan, 1997; Pier et al., 1985), and recovery from protein malnutrition is delayed by exposure to aflatoxin (Adhikari et al., 1994; Rogers, 1993).

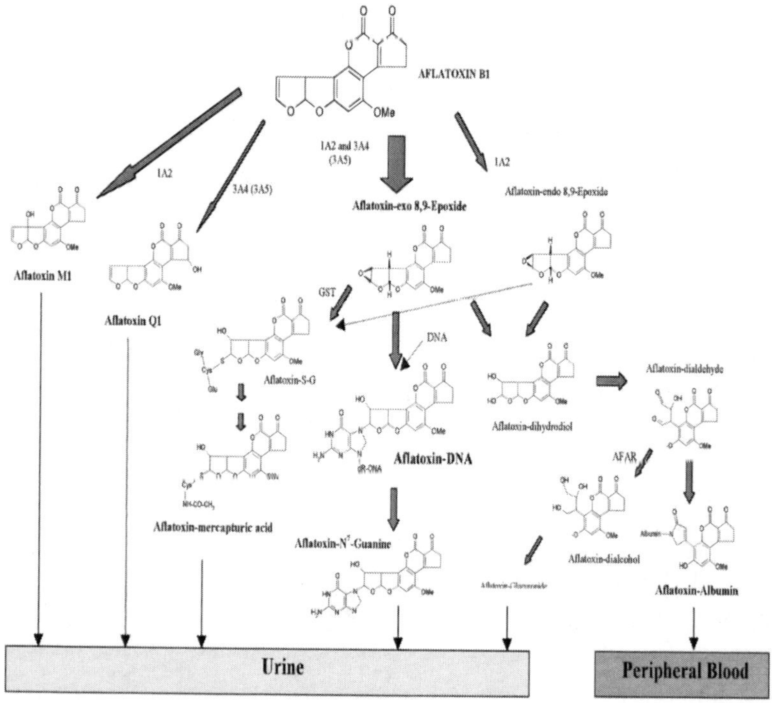

Figure 2. Principle metabolism of aflatoxin B1 leading to reactive metabolites and biomarkers. 1A2, CYP1A2; 3A4, CYP3A4; 3A5, CYP3A5; GST, glutathione S-transferase; AFAR, aflatoxin aldehyde reductase; Aflatoxin-S-G, aflatoxin-glutathione conjugate. (Wild and Turner, 2002, Mutagenesis 17(6), Figure 1, pg 472).

Biomarker-based methodologies may be employed to measure human exposure to food contaminants. It has the advantage of being directly relevant to human risk and potentially identifies the hazard, to give early warning by signaling the early effects of exposure and increased susceptibility, thus allowing opportunities to arrest a disease through timely intervention (Perera, 1996). Human exposure to AFB1 using biomarkers approach was reported in the late 1980s and was one of the first applications of biomarkers to human exposure to food chemicals or contaminants (Gan et al., 1988; Zhu et al., 1987). Statistically significant correlations between the biomarker aflatoxin-albumin and urinary excretion of the metabolite AFM1 were reported in some studies (Gan et al., 1988; Piekkola et al., 2012). Subsequent researches have tended to investigate AFB1 bound to serum or plasma albumin (Turner et al., 1998; Wild et al., 1992) and urinary AFB1-DNA adducts (Groopman et al., 1992b), wherein a strong correlation coefficient of 0.80 with dietary exposure was found. Comparison of intakes by the biomarker approach with those obtained from consumption data has been undertaken (European Commission, 1997). The level of agreement varies where the plasma-derived intake estimate was lower by a factor of between 5-10, for two of the five data sets.

DETERMINATION OF AFLATOXIN METABOLITES AS BIOMARKERS IN HUMAN

Epidemiological studies on aflatoxins exposure were extensively studied during 1980s and early 1990s; therefore, some studies cited in this section are relatively dated. The development of molecular and biochemical epidemiology have led analysis of aflatoxins extend beyond the determination of levels in foods and feeds. Thus, the measurement of actual ingestion of aflatoxins in human populations and the biologically effective dose of aflatoxins, i.e., the amount of the activated compound that has actually reacted with cellular targets such as DNA, protein or RNA is now feasible with the use of biomarkers. Several molecular biomarkers have been developed and applied, including the AFB1 metabolites and AFB1 macromolecular adducts. These involved AFM1 and AFB1-N^7-guanine in urine as well as AFB1-albumin adducts in serum (Wang et al., 2001). The latter adduct is considered to reflect more the longer term intake of AFB1 based on the longer half-life of albumin in humans compared to urinary metabolites (Groopman et al., 1994). More stable and less fluctuation for aflatoxin-albumin adduct than urinary excretion of aflatoxin metabolites over a same period of time is observed (Groopman et al., 1992a; Wild et al., 1992). This is why it is considered to have greater value as a biomarker. In fact, the aflatoxin B1-lysine adduct, which is isolated from serum albumin, is a validated biomarker of aflatoxins exposure and risk (Chen et al., 2013; Jolly et al., 2015; Sabbioni et al., 1990; Turner et al., 2005). It has thus been widely applied in epidemiological studies and provides a considerable database for human aflatoxins exposure in different countries (Wild et al., 1990a, 1993). Aflatoxin-albumin adduct has been analyzed in a variety of recent studies, for instance, in African countries, e.g., Gambia (Castelino et al., 2014, 2015), Tanzania (Routledge et al., 2014; Shirima et al., 2013), and Kenya (Yard et al., 2013), as well as in Asian countries such as China (Chen et al., 2013), Nepal (Groopman et al., 2014), India (Anitha et al., 2014), and Bangladesh (Groopman et al., 2014).

Biomarker based exposure estimates are more reliable and reflect not only the dietary exposure of the individual but also the uptake, toxicokinetics and toxicodynamics (Ediage et al., 2012). Methods that have been employed include immunological assays using specific antibodies or antisera such as enzyme-linked immunosorbent assay (ELISA), radio-immunoassay (RIA), chromatography techniques like thin layer chromatography (TLC), liquid chromatography with fluorescence or mass spectrometric detection (HPLC-FLD, HPLC-MS/MS) with various extraction and clean-up methods (Kensler et al., 1998; Capriotti et al., 2012). RIA is comparably less common than other techniques as it requires radio-labeled substrates, access to scintillation counters, and skilled personnel. However, multi-biomarker studies are more favored in recent time to characterize the concentrations of various mycotoxins in exposed persons, as well as in the general population (Warth et al., 2013). Gerding et al. (2014), Ezekiel et al., (2014), Shirima et al., (2013) and Ediage et al. (2012) report some examples of studies using multi-biomarker approach.

Human Urine

Various aflatoxins and their levels in human urine have been reviewed (Autrup and Autrup, 1992; Fromme et al. 2016; Yourtee and Kirk-Yourtee, 1989). Aflatoxin B1, B2, G1, G2, M1, Q1, P1, B2a, aflatoxin-N^7-guanine and aflatoxicol have been measured in human urine. Nevertheless, the levels of these urinary metabolites or adducts only reflect intake on the previous day due to the short half-life of such metabolites in the urine. The nucleic acid adduct aflatoxin-N^7-guanine were focused since it reflected DNA damage in the presumed target cell for aflatoxins carcinogenesis (Groopman et al., 1993).

According to Zhu et al. (1987), just around 1.2% to 2.2% of aflatoxin present in the diet is excreted as AFM1 in human urine. Several studies on human exposure to aflatoxins using urinary AFM1 as biomarkers have been reported. Generally, studies in recent decades show primarily low levels of urinary AFM1. However, the findings vary with geographic factors and appear distinctly higher in some individuals whose staple food are potentially contaminated by aflatoxins. Lei et al., (2013) reported the detection of AFM1 in 84% of 600 samples with a geometric mean of 50.3 ng/l in pregnant women in China's Zhejiang province. A bigger percentage of higher levels were found in people from rural areas compared to those in urban regions. Ali et al. (2016) reported that higher mean level of urinary AFM1 was detected in winter (27.7 ± 42.6 pg/ml) than in summer (13.6 ± 21.2 pg/ml) and differences were also observed between rural and the urban areas in Bangladesh. In Germany, none of the targeted aflatoxins (AFB1, AFB2, AFG2 and AFM1) was detected in 101 samples among German adults (Gerding et al., 2014). 78% (n = 54) of the urine samples contained >0.6 pg/ml of AFM1 in Brazilian population (Romero et al.,

2010). The highest levels ever found was reported in a study carried out in Sierra Leone, Africa, where concentration up to 374 ng/ml was detected among 244 primary school children (Jonsyn-Ellis, 2000).

Detection of the presence of aflatoxins, particularly AFM1, in human urine involves TLC, ELISA, HPLC with fluorescence detection and LC-MS/MS. TLC and HPLC have been adopted in the study by Nyathi et al. (1987) to analyse AFM1 excretion into over 1200 urine samples. Wild et al., (1986, 1992) extracted aflatoxins from urine using immunopurification and quantitated using polyclonal antibody in a highly sensitive ELISA. More recently, ELISA kit with coated anti-AFM1 antibody was used by Lei et al. (2013) to analyze AFM1 in 600 urine samples collected from volunteers from Zhejiang province in China. In an Egyptian study, besides single analyte of AFM1, determination of total and individual aflatoxin B1, B2, G1, G2, M1, M2, B2a, G2a, B3, GM1, P as well as aflatoxicol in urine in infants was performed by using a two-dimensional TLC (Hatem et al., 2005). The use of immunoaffinity clean-up and detection by HPLC with fluorescence detector was reportedly used for determining the presence of AFM1 in urine samples for the Brazil (Romero et al., 2010) and Bangladeshi population (Ali et al., 2016) and for the analysis of urinary aflatoxin metabolites (AFB1, AFB2, AFG1, AFG2 and AFM1) in young children in Egypt and Guinea (Polychronaki et al., 2008). Lately, the HPLC has been substituted by LC-MS/MS to determine mycotoxins exposure using the multi-biomarker approach.

Human Blood

Detection of free aflatoxins was focused initially (Hendrickse et al., 1982; Tsuboi et al., 1984) but subsequent identification of metabolic rate of aflatoxins, aflatoxin-albumin adduct is preferred as it is able to integrate exposure over a period of months based on the half-life of albumin in humans (Sabbioni et al. 1987). Aflatoxin that binds to albumin can be AFB1 and AFG1 but not AFB2 and AFG2 which can be metabolised to 8, 9-epoxide (Egal et al., 2005). However, the aflatoxin-albumin adduct levels should be regarded as a measure of the levels of AFB1 ingested, as the presence of AFG1 in a contaminated food is rare.

Albumin is the only protein in serum which binds AFB1 to form a high level of adducts in rats (Skipper et al., 1985) while haemoglobin binds AFB1 in a very low yield (Tannenbaum and Skipper, 1984). Albumin is readily extracted from human blood and it provides a relatively non-intrusive measure of biologically effective dose of ingested AFB1 (Wild et al., 1996). This approach may be useful for rapid screening of samples for acute exposures and also reflects chronic exposure that is not available from other markers such as the aflatoxin-N^7-guanine adduct in urine.

The major analytical techniques currently available for measuring AFB1-albumin adduct levels in human blood use ELISA, RIA, IAC followed by HPLC with fluorescence detection and LC-MS/MS techniques (Gan et al., 1988; Sabbioni et al., 1990; Wild et al., 1990b; Chapot and Wild, 1991; Wang et al., 1996; Walton et al., 2001; McCoy et al., 2005; Shuaib et al., 2010; Xu et al., 2010, Shirima et al., 2013). AFB1-lysine is the only AFB1 adduct which has been structurally identified in enzymatically digested plasma albumin (Guengerich et al., 2002). This led to the development of a specific monoclonal antibody (IIA4B3) for AFB1-lysine adduct using a synthetic AFB1-lysine-cationized bovine serum albumin conjugate (Wang et al., 2001). Previous studies had either used a polyclonal antiserum-based ELISA (Wild et al., 1990b; Chapot and Wild, 1991) or a monoclonal antibody 2B11-based RIA (Gan et al., 1988; Wang et al., 1996) to measure AFB1-albumin adducts. The IIA4B3 was found about 12-fold more sensitive than 2B11 for binding to AFB1-lysine when ^{3}H-AFB1-lysine was used as the tracer (Wang et al., 2001). In addition, a more specific, precise and sensitive detection of AFB1-lysine with D4-AFB1-lysine as the internal standard using an isotope dilution liquid chromatography mass spectrometry (IDMS) has also been developed by McCoy et al. (2005). The measured adduct concentration is expressed as aflatoxin equivalents per mg albumin.

Highly correlated data generated from various methods, for example, ELISA and HPLC-fluorescence (Wild et al., 1990a), RIA and HPLC-fluorescence (Wang et al., 1996) as well as ELISA and LC-MS/MS (Scholl et al., 2006) have been reported. However, aflatoxin levels measured by antibody-based methods are usually higher than those measured using the LC/MS method when isotope internal standard was coupled because the ELISA may be measuring other aflatoxin adducts in addition to the target adduct.

Aflatoxin-albumin was used to measure aflatoxin exposure in West Africa countries including Gambia, Guinea, Ghana, Benin, Togo and Tanzania (Wild et al., 2000; Gong et al., 2002, 2003; Egal et al., 2005; Jolly et al., 2006; Shirima et al., 2013), in China (Ross et al., 1992; Qian et al., 1994), and infectious disease linked immune suppression (Turner et al., 2003; Jiang et al., 2005). Further, aflatoxin-albumin is regularly used as the surrogate efficacy biomarker of AFB1 exposure for the assessment of different therapeutic/intervention agents and techniques in human intervention trials (Kensler et al., 1998).

Breast Milk

Aflatoxin metabolites can also be found in lactating animals and therefore in human breast milk. Fromme et al. (2016) and Cherkani-Hassani et al. (2016) reviewed the levels of AFM1 in breast milk from several studies. The most common method used to quantify the concentrations of mycotoxins in breast milk is HPLC; about 25% by ELISA or both simultaneously (Cherkani-Hassani et al., 2016). Studies on occurrence of AFM1 in breast milk in Egypt (Polychronaki et al., 2007), Iran (Afshar et al., 2013; Sadeghi et al., 2009), Australia, Thailand (El-Nezami et al., 1995), Italy (Turconi et al., 2004) Turkey (Gurbay et al., 2010), and Nigeria (Adejumo et al., 2013) are documented.

Other Human Fluids and Tissues

Although AFB1 is a potent liver toxin it can also affect other tissues such as the lungs, kidney and colon, in both rodents and humans (Larsson and Tjalve, 1995). Aflatoxins were detected in nasal secretions, sputum and tissue biopsies (lung, liver, brain) from patients exposed to toxin producing molds in their environment (Hooper et al., 2009). Additionally, feces has been proposed as a marker for aflatoxins exposure since they are an important route of excretion of both the unabsorbed AFB1 and metabolites formed from the absorbed toxin (Mykkänen et al., 2005). Excretion of AFQ1 and AFM1 in human feces has been quantified in 83 young Chinese males selected from 300 subjects based on detectable urinary AFM1 in Guangzhou, southern China. The study shows that fecal AFQ1 was 60 times higher than that of AFM1.

Risk Assessment of Aflatoxins

Risk assessment is a four-step process (hazard identification, hazard characterization, exposure assessment and risk characterization) (Barlow et al., 2002; Renwick et al., 2003), that determines the possible adverse effects of a specific contaminant such as aflatoxins on an organism. Exposure assessment takes into consideration the food consumption data with value on the concentrations of aflatoxins in the diet. The estimated dietary intake from exposure assessment could be determined to reduce human risks (Huong et al., 2016). Expert committees and institutions, e.g., the European Food Safety Authority, evaluates the toxicity of certain contaminants after considering relevant scientific information and sets the toxicological values, such as the tolerable daily intake (TDI), maximum tolerable daily intake (PMTDI) at the end of a risk assessment process. The TDI characterizes the total amount of a substance that can be taken daily over a lifetime without a considerable health risk. Some TDI or risk values related to mycotoxins have been reviewed (Fromme et al., 2016). However, in the case of AFB1, no TDI was established for this genotoxic carcinogen (EFSA, 2006). One of the common methods of assessing the carcinogenic and genotoxic mycotoxins is the estimation of the margin of exposure (MOE) (EFSA, 2006; JECFA, 2007). The MOE is simply a ratio, obtainable by dividing the reference point on the dose that causes a defined impact (benchmark dose, BMD) in animals, by the estimated human intake of AFB1. A MOE is not a quantification of risk, however MOE <10,000 can be considered a high risk to public health. Application of MOE approach to the risk assessment of aflatoxins in dried fruits (Heshmati et al., 2017), unrecorded beers (Okaru et al., 2017), pasta and bakery products (Bol et al., 2016), nuts and nut products (Leong et al., 2011) have been reported.

ACKNOWLEDGEMENTS

The authors would like to thank Chuah Lucie for editing and proofreading the text.

REFERENCES

Abdullah, N., Nawawi, A., Othman, I., (1998). Survey of fungal counts and natural occurrence of aflatoxins in Malaysian starch-based foods. *Mycopathologia, 143*, 53-58.

Adejumo, O., Atanda, O., Riola, A., Somorin, Y., Bandyopadhyay, R., Ritieni, A., (2013). Correlation between aflatoxin M1 content of breast milk, dietary exposure to aflatoxin B1 and socioeconomic status of lactating mothers in Ogun State, Nigeria. *Food and Chemical Toxicology,* 56, 171–177.

Adhikari, M., Ramjee, G., Berjak, P., (1994). Aflatoxin, kwashiorkor and morbidity. *Journal of Natural Toxins*, 2, 1-3.

Afshar, P., Shokrzadeh, M., Kalhori, S., Babaee, Z., Saeedi Saravi, S. S., (2013). Occurrence of Ocharatoxin A and Aflatoxin M1 in human breastmilk in Sari, Iran. *Food Control*, 31, 525–529.

Ali, N., Hashim, N. H., Saad, B., Safan, K., Nakajima, M., Yoshizawa, T., (2005). Evaluation of a method to determine the natural occurrence of aflatoxins in commercial traditional herbal medicines from Malaysia and Indonesia. *Food and Chemical Toxicology,* 43, 1763-1772.

Ali, N., Hashim, N. H., Yoshizawa, T., (1999). Evaluation and application of a simple and rapid method for the analysis of aflatoxins in commercial foods from Malaysia and the Philippines. *Food Additives and Contaminants,* 16(7), 273-280.

Anitha, S., Raghunadharao, D., Waliyar, F., Sudini, H., Parveen, M., Rao, R., Kumar, P. L., (2014). The association between exposure to aflatoxin, mutation in TP53,infection with hepatitis B virus, and occurrence of liver disease in a selected population in Hyderabad, India. *Mutation Research - Genetic Toxicology and Environmental Mutagenesis,* 15, 23–28.

Autrup, H., Autrup, J. L., (1992). In: D. Bhatnagar, E. B. Lillehoj and D. K. Arora, (eds). *Handbook of applied mycology,* volume 5. New York: Marcel Dekker. P. 213.

Ayub, M. Y., Sachan, D. S., (1997). Dietary factors affecting aflatoxin B1 carcinogenicity. *Malaysian Journal of Nutrition,* 3, 161-179.

Azziz-Baumgartner, E., Lindblade, K., Gieseker, K., Rogers, H. S., Kieszak, S., Njapau, H., Schleicher, R., McCoy, L. F., Misore, A., DeCock, K., Rubin, C., Slutsker, L., and the Aflatoxin Investigative Group, (2005). Case-control study of an acute aflatoxicosis outbreak, Kenya, 2004. *Environmental Health Perspectives,* 113(12), 1779-1783.

Barlow, S. M., Grieg, J. B., Bridges, J. W., Carere, A., Carpy, A. J. M., Galli, C. L., Kleiner, J., Knudsen, I., Koeter, H. B. W. M., Levy, L. S., Madsen, C., Mayer, S., Narbonne, J. F., Pfannbuch, F., Prodanchuk, M. G., Smith, M. R., Steinberg, P., (2002). Hazard identification by methods of animal-based toxicology. *Food and Chemical Toxicology,* 40, 145-191.

Battilani, P., Toscano, P., Van der Fels-Klerx, H. J., Moretti, A., Camardo Leggieri, M., Brera, C., Rortais, A., Goumperis, T., Robinson, T., (2016). Aflatoxin B1 contamination in maize in Europe increases due to climate change. *Scientific Reports,* 6, 24328.

Benammar, L., Menasria, T., Chergui, A., Benfiala, S., (2017). Indoor fungal contamination of traditional public baths (Hammams). *International biodeterioration and Biodegradation*, 117, 115-122.

Bennett, J. W., Klich, M., (2003). Mycotoxins. *Clinical Microbiology Reviews,* 16, 497-516.

Bhat, R., Rai, R. V., Karim, A. A., (2010). Mycotoxins in food and feed: present status and future concerns. *Comprehensive Reviews in Food Science and Food Safety*, 9, 57-81.

Bhat, R. V., Vasanthi, S., (2003). Food Safety in Food Security and Food Trade. Mycotoxin food safety risk in developing countries. International Food Policy Research Institute, Focus 10, Brief 3 of 17.

Bol, E. K., Araujo, L., Veras, F. F., Welke, J. E., (2016). Estimated exposure to zearalenone, ochratoxin A and aflatoxin B1 through the consume of bakery products and pasta considering effects of food processing. *Food and Chemical Toxicology,* 89, 85-91

Bonomi, A., Quarantelli, A., Mazzali, I., Cabassi, E., Corradi, A., Lecce, R., Ubaldi, A., Fusari, A. and Chizzolini, A., (1994). Effects of aflatoxin B1 contaminated rations on the productive efficiency and on the meat yield and quality in fattening pigs (experimental contribution). *Rivista di Scienza dell'Alimentazione*, 22, 351-377.

Boonen, J., Malysheva, S. V., Taevernier, L., Diana Di Mavungu, J., De Saeger, S., De Spiegeleer, B., (2012). Human skin penetration of selected model mycotoxins. *Toxicology,* 301, 21–32.

Bourgeois, C., Olson, L., Comer, D., Evans, H., Keschamras, N., Cotton, R, Grossman, R. and Smith, T., (1971). Encephalopathy and fatty degeneration of the viscera: a clinic-pathologic analysis of 40 cases. *American Journal of Clinical Pathology,* 56, 558-571.

Cano-Sancho, G., Perello, G., Nadal, M., Domingo, J. L., (2015). Comparison of the nutritional composition and the concentrations of various contaminants in branded and private label yogurts. *Journal of Food Composition and Analysis,* 42, 71-77.

Capriotti, A. L., Caruso, G., Cavaliere, C., Foglia, P., Samperi, R., Lagana, A., (2012). Multiclass mycotoxin analysis in food, environment and biological matriceswith chromatography/mass spectrometry. *Mass Spectrometry Reviews,* 31, 466–503.

Castelino, J. M., Dominguez-Salas, P., Routledge, M. N., Prentice, A. M., Moore, S. E., Hennig, B. J., Wild, C. P., Gong, Y. Y., (2014). Seasonal and gestation stageassociated differences in aflatoxin exposure in pregnant Gambian women. *Tropical Medicine and International Health,* 19, 348–354.

Castelino, J. M., Routledge, M. N., Wilson, S., Dunne, D. W., Mwatha, J. K., Gachuhi, K., Wild, C. P., Gong, Y. Y., (2015). Aflatoxin exposure is inversely associated withIGF1 and IGFBP3 levels in vitro and in Kenyan schoolchildren. *Molecular Nutrition and Food Research,* 59, 574–581.

Chao, T. C., Maxwell, S. M., Wong, S. Y., (1991). An outbreak of aflatoxicosis and boric acid poisoning in Malaysia: a clinicopathological study. *Journal of Pathology*, 164, 225-233.

Chapot, B., Wild, C. P., (1991). ELISA for quantification of aflatoxin-albumin adducts and their application to human exposure assessment. In: M. Warthol, D. van Velzer and G. Bullock, (eds.). *Techniques in diagnostic pathology.* London: Academic Press. P. 135-155.

Cheeke, P. R., Shull, L. R., (1985). Natural Toxicants in Feeds and Poisonous Plants. Westport: AVI Publisher. P. 402-421.

Chen, J. G., Egner, P. A., Ng, D., Jacobson, L. P., Munoz, A., Zhu, Y. R., Qian, G. S., Yuan, J. M., Groopman, J. D., Kensler, T. W., (2013). Reduced aflatoxin exposure presages decline in liver cancer mortality in an endemic region of China. *Cancer Prevention.Research,* 6, 1038–1045.

Cherkani-Hassani A, Mojemmi B, Mouane N., (2016). Occurrence and levels of mycotoxins and their metabolites in human breast milk associated to dietary habits and other factors: A systematic literature review, 1984-2015. *Trends in Food Science and Technology*, 50, 56-69.

Degen, G. H., (2011) Tools for investigating workplace-related risks from mycotoxin exposure. *World Mycotoxin Journal,* 4, 315–327.

de Onis, M., Branca, F., (2016). Childhood stunting: a global perspective. *Maternal and Child Nutrition.* 1(12 Suppl), 12-26.

de Ruyck, K., De Boevre, M., Huybrechts, I., De Saeger, S., (2015). Dietary mycotoxins, co-exposure, and carcinogenesis in humans: Short review. *Mutation Research-Review in Mutation Research,* 766, 32-41.

Eaton, D. L., Gallagher, E. P., (1994). Mechanisms of aflatoxin carcinogenesis. *Annual Review of Pharmacology and Toxicology*, 34, 135-172.

Eaton, D. L., Ramsdell, H. S., Neal, G., (1993). Biotransformation of aflatoxins. In: D. L. Eaton and J. D. Groopman, (eds). The toxicology of aflatoxins: human health, veterinary, and agricultural significance. London: Academic Press. P. 45-72.

Ediage, E. N., Di Mavungu, J. D., Song, S., Wu, A., Van Peteghem, C., De Saeger, S., (2012). A direct assessment of mycotoxin biomarkers in human urine samples by liquid chromatography tandem mass spectrometry. *Analytica Chimica Acta,* 741, 58-69.

EFSA, (2006). Opinion of the scientific panel on contaminants in the food chain on a request from the commission related to ochratoxin A in food. The EFSA Journal. [Online]. 365, 1e56. [Accessed 5 May 2017]. Available from World Wide Web: https://www.efsa.europa.eu/en/efsajournal/pub/365.

EFSA, (2007). Opinion of the Scientific Panel on Contaminants in the Food Chain on a request from the Commission related to the potential increase of consumer health risk by a possible increase of the existing maximum levels for aflatoxins in almonds, hazelnuts and pistachios and derived products. *The EFSA Journal. [Online].* 446, [Accessed 5 May 2017]. p. 1-127. Available from World Wide Web: http://www.efsa.europa.eu/en/scdocs/scdoc/446.htm.

Egal, S., Hounsa, A., Gong, Y. Y., Turner, P. C., Wild, C. P., Hall, A. J., Hell, K., Cardwell, K. F., (2005). Dietary exposure to aflatoxin from maize and groundnut in young children from Benin and Togo, West Africa. *International Journal of Food Microbiology,* 104, 215-224.

Ellis, W. O., Smith, J. P., Oldham, J. H., (1991). Aflatoxins in food: occurrence, biosynthesis, effects on organism's detection and methods of control. *Critical Reviews in Food Science and Nutrition,* 30, 403-439.

El-Nezami, H. S., Nicoletti, G., Neal, G. E., Donohue, D. C., Ahokas, J. T., (1995). Aflatoxin M1 in human breast milk samples from Victoria, Australia and Thailand. *Food and Chemical Toxicolology*, 33, 173–179.

Er, B., Demirhan, B., Yentur, G., (2014). Short communication: investigation of aflatoxin M1 levels in infant follow-on milks and infant formulas sold in the markets of Ankara, *Turkey. Journal of Dairy Science,* 97(6), 3328-3331.

European Commission, (1997). Improvement of Knowledge of Food Consumption with a View to Protection of Public Health by Means of Exchanges and Collaboration Between Database Managers (Report of experts participating in Task 4.1). Luxembourg: Office for Official Publications of the European Commission.

Ezekiel, C. N., Warth, B., Ogara, I. M., Abia, W. A., Ezekiel, V. C., Atehnkeng, J., Sulyok, M., Turner, P. C., Tayo, G. O., Krska, R., Bandyopadhyayet, R., (2014). Mycotoxin exposure in rural residents in northern Nigeria: a pilot study using multi-urinary biomarkers. *Environment International,* 66, 138-145.

Farah Nadira, A., Rosita, J., Norhaizan, M. E., Mohd Redzwan, S., (2017). Screening of aflatoxin M1 occurrence in selected milk and dairy products in Terengganu, Malaysia. *Food Control*, 73, 209-214.

Fromme, H., Gareis, M., Völkel, W., Gottschalk, C., (2016). Overall internal exposure to mycotoxins and their occurrence in occupational and residential settings – An overview. *International Journal of Hygiene and Environmental Health*, 219, 143–165.

Fung, F., Clark, R. F., (2004). Health effects of mycotoxins: a toxicological overview. *Journal of Toxicology,* 42, 217-234.

Gan, L. S., Skipper, P. L., Peng, X. C., Groopman, J. D., Chen, J. S., Wogan, G. N., Tannenbaum, S. R., (1988). Serum albumin adducts in the molecular epidemiology of aflatoxin carcinogenesis: correlation with aflatoxin B1 intake and urinary excretion of aflatoxin M1. *Carcinogenesis,* 9, 1323-1325.

Gerding, J., Cramer, B., Humpf, H. U., (2014). Determination of mycotoxin exposure in Germany using an LC–MS/MS multibiomarker approach. *Molecular Nutrition and Food Research,* 58, 2358–2368.

Gong, Y. Y., Cardwell, K., Hounsa, A., Egal, S., Turner, P. C., Hall, A. J., Wild, C. P., (2002). Dietary aflatoxin exposure and impaired growth in young children from Benin and Togo: cross sectional study. *British Medical Journal,* 325, 20-21.

Gong, Y. Y., Egal, S., Hounsa, A., Turner, P. C., Hall, A. J., Cardwell, K. F., Wild, C. P., (2003). Determinants of aflatoxin exposure in young children from Benin and Togo, West Africa: the critical role of weaning. *International Journal of Epidemiology,* 32, 556-562.

Groopman, J. D., Egner, P. A., Schulze, K. J., Wu, L. S., Merrill, R., Mehra, S., Shamim, A. A., Ali, H., Shaikh, S., Gernand, A., Khatry, S. K., LeClerq, S. C., West Jr., K. P., Christian, P., (2014). Aflatoxin exposure during the first 1000 days of life in rural South Asia assessed by aflatoxin B(1)-lysine albumin biomarkers. *Food and Chemical Toxicology.* 74, 184-189.

Groopman, J. D., Hall, A. J., Whittle, H., Hudson, G. J., Wogan, G. N., Montesano, R., Wild, C. P., (1992a). Molecular dosimetry of aflatoxin-N^7-guanine in human urine obtained in The Gambia, West Africa. *Cancer Epidemiology, Biomarkers and Prevention,* 1, 221-227.

Groopman, J. D., Wild, C. P., Hasler, J., Junshi, C., Wogan, G. N., Kensler, T. W., (1993). Molecular epidemiology of aflatoxin exposures: validation of aflatoxin-N^7-guanine levels in urine as a biomarkers in experimental rat models and human. *Environmental Health Perspectives,* 99, 107-113.

Groopman, J. D., Wogan, G. N., Roebuck, B. D., Kensler, T. W., (1994). Molecular biomarkers for aflatoxins and their applications to human cancer prevention. *Cancer Research,* 54, 1907S-1911S.

Groopman, J. D., Zhu, J. Q., Donahue, P. R., Pikul, A., Zhang, L. S., Chen, J. S., Wogan, G. N., (1992b). Molecular dosimetry of urinary aflatoxin-DNA adducts in people living in Guangxi Autonomous Region, People's Republic of China. *Cancer Research,* 52, 45-52.

Guengerich, F. P., Arneson, K. O., Williams, K. M., Deng, Z., Harris, T. M., (2002). Reaction of aflatoxin B1 oxidation products with lysine. *Chemical Research in Toxicology,* 15, 780-792.

Guengerich, F. P., Johnson, W. W., Shimada, T., Ueng, Y. F., Yamazaki, H., Langouet, S., (1998). Activation and detoxication of aflatoxin B1. *Mutation Research*, 402, 121-128.

Gürbay, A., Sabuncuogluo, S. A., Girgin, G., Sahin, G., Yigit, S., Yurdakök, M., Tekinalp, G., (2010). Exposure of newborns to aflatoxin M1 and B1 from mothers' breast milk in Ankara, Turkey. *Food and Chemical Toxicology,* 48, 314–319.

Haas, D., Habib, J., Galler, H., Buzina, W., Schlacher, R., Marth, E., Reinthaler, F. F., (2007). Assessment of indoor air in Austrian apartments with and without visible mold growth. *Atmospheric Environment,* 41, 5192-5201.

Hashemi, M., (2016). A survey of aflatoxin M1 in cow milk in Southern Iran. *Journal of Food and Drug Analysis,* 24, 888-893.

Hatem, N. L., Hassab, H. M. A., Al-Rahman, E. M. A., El-Deeb, S. A., (2005). Prevalence of aflatoxins in blood and urine in Egyptian infants with protein-energy malnutrition. *Food and Nutrition Bulletin*, 26(1), 49-56.

Hedayati, M. T., Mayahi, S., Denning, D. W., (2010). A study on Aspergillus species in houses of asthmatic patients from Sari City, Iran and a brief review of the health effects of exposure to indoor Aspergillus. *Environmental Monitoring and Assessment*, 168, 481-487.

Hedayati, M. T., Mayahi, S., Movahedi, M., Shokohi, T., (2011). Study on fungal flora of tap water as a potential reservoir of fungi in hospitals in Sari city, Iran. *Journal of Medical Mycology,* 21, 10-14.

Hendrickse, R. G., Coulter, J., Lamplugh, S., MacFarlane, S., Williams, T., Omer, M., Suliman, G., (1982). Aflatoxins and kwashiorkor: a study in Sudanese children. *British Medical Journal,* 285, 843-846.

Heshmati, A., Zohrevand, T., Khaneghah, A. M., Nejad, A. S. M., Sant'Ana, A. S., (2017). Co-occurrence of aflatoxins and ochratoxin A in dried fruits in Iran: Dietary exposure risk assessment. *Food and Chemical Toxicology* 106, 202-208.

Hooper, D. G., Bolton, V. E., Guilford, F. T., Straus, D. C., (2009). Mycotoxin detection in human samples from patients exposed to environmental molds. *International Journal of Molecular Sciences,* 10, 1465-1475.

Hsieh, D., (1988). Potential human health hazards of mycotoxins. In: S. Natori, K. Hashimoto, and Y. Ueno (eds.). Mycotoxins and phytotoxins. Third Joint Food and Agriculture Organization/WHO/ United Nation/Program International Conference of Mycotoxins, Amsterdam, The Netherlands. P. 69-80.

Huong, B. T. M., Tuyen, L. D., Tuan, D. H., Brimer, L., Dalsgaard, A., (2016). Dietary exposure to aflatoxin B1, ochratoxin A and fuminisins of adults in Lao Cai province, Viet Nam: a total dietary study approach. *Food and Chemical Toxicology,* 98, 127-133.

Hussein, H. S., Brasel, J. M., (2001). Toxicity, metabolism, and impact of mycotoxins on humans and animals. *Toxicology,* 167, 101-134.

IARC, (1993). Some naturally occurring substances: Food items and constituents, heterocyclic aromatic amines and mycotoxins. IARC monograph on the evaluation of carcinogenic risk to humans. [Online]. 56, [Accessed 22 April 2010], p. 489-521. Available from World Wide Web: http://monographs.iarc.fr/ENG/Monographs/vol56/ volume56.pdf.

IARC, (2002). Some traditional herbal medicines, some mycotoxins, naphthalene and styrene: Summary of data reported and evaluation. IARC monograph on the evaluation of carcinogenic risk to humans. [Online]. 82, [Accessed 22 May 2017], p. 1-556. Available from World Wide Web: http://monographs.iarc.fr/ENG/Monographs/ vol82/mono82.pdf.

IARC, (2012). Monographs on the evaluation of carcinogenic risks to humans: chemical agents and related occupations. A review of human carcinogens. Lyon, France: International Agency for Research on Cancer 100F, 224–248. [Online]. 56, [Accessed 22 May 2017], p. 489-521. Available from World Wide Web: http://monographs.iarc.fr/ENG/ Monographs/vol100F/mono100F.pdf.

Iqbal, S. Z., Jinap, S., Pirouz, A. A., Ahmad Faizal, A. R., (2015). Aflatoxin M1 in milk and dairy products, occurrence and recent challenges: A review. *Trends in Food Science & Technology,* 46, 110-119.

Jakšić, D., Šegvić Klarić M., (2014). A year-round investigation of indoor airborne fungi in Croatia. *Archives of Industrial Hygiene and Toxicology,* 65, 209-218.

JECFA, (2007). Evaluation of Certain Food Additives and Contaminants: Sixty-eighth Report of the Joint FAO/WHO Expert Committee on Food Additive. WHO Technical Report Series No 947. WHO, Geneva, Switzerland. [Online]. [Accessed 22 May 2017], p. 169-180. Available from World Wide Web: http://apps.who.int/iris/bitstream/10665/43870/1/9789241209472_eng.pdf.

Jiang, Y., Jolly, P. E., Ellis, W. O., Wang, J. S., Phillips, T. D., Williams, J. H., (2005). Aflatoxin B1 albumin adduct levels and cellular immune status in Ghanaians. *International Immunology,* 17, 807-814.

Jolly, P. E., Akinyemiju, T. F., Jha, M., Aban, I., Gonzalez-Falero, A., Joseph, D., (2015). Temporal variation and association of aflatoxin B1 albumin-adduct levels with socio-economic and food consumption factors in HIV positive adults. *Toxins.* 7, 5129-5140.

Jolly, P., Jiang, Y., Ellis, W., Awuah, R., Nnedu, O., Phillips, T., Wang, J. S., Afriyie-Gyawu, E., Tang, L., Person, S., Williams, J., Jolly, C., (2006). Determinants of aflatoxin levels in Ghanaians: sociodemographic factors, knowledge of aflatoxin and food handling and consumption practices. *International Journal of Hygiene and Environmental Health,* 209(4), 345-358.

Jonsyn-Ellis, F., (2000). Seasonal variation in exposure frequency and concentration levels of aflatoxins and ochratoxins in urine samples of boys and girls. *Mycopathologia,* 152, 35-40.

Kelman, B. J., Robbins, C. A., Swenson, L. J., Hardin, B. D., (2004). Risk from inhaled mycotoxins in indoor office and residential environments. *International of Journal Toxicology,* 23, 3–10.

Kensler, T. W., Groopman, J. D., Roebuck, B. D., (1998). Use of aflatoxin adducts as intermediate endpoints to assess the efficacy of chemopreventive interventions in animals and man. *Mutation Research,* 402, 165-172.

Khan, A. A. H., Karuppayil, S. M., Manoharachary, C., Kunwar, I. K., Waghray, S., (2009). Isolation, identification and testing for allergenicity of fungi from air-conditioned indoor environments. *Aerobiologia,* 25, 119-123.

Kilburn, K. H., (2004). Role of molds and mycotoxins in being sick in buildings: neurobehavioral and pulmonary impairment. *Advances in Applied Microbiology,* 55, 339–359.

Klich, M., Mullaney, E., Daly, C., Cary, J., (2000). Molecular and physiological aspects of aflatoxin and sterigmatocystin biosynthesis by Aspergillus tamariiand A. ochraceoroseus. *Applied Microbiology and Biotechnology,* 53, 605–609.

Kos, J., Levi, J., Duragi, O., Koki, B., Miladinovi, I., (2014). Occurrence and estimation of aflatoxin M1 exposure in milk in Serbia. *Food Control,* 38:41-46.

Krishnamachari, K. A., Nagaarajan, V., Bhat, R. V. and Tilak, T. B., (1975). Hepatitis due to aflatoxicosis – an outbreak in Western India. *The Lancet* 305, 1061-1063.

Kurtzman, C., Horn, B., Hesseltine, C., (1987). Aspergillus nomius, a new aflatoxin-producing species related to Aspergillus flavus and Aspergillus tamarii. *Antonie Van Leeuwenhoek,* 53, 147–158.

Larsson, P., Tjalve, H., (1995). Extrahepatic bioactivation of aflatoxin B1 in fetal, infant and adult rats. *Chemico-Biological Interactions,* 94(1), 1-19.

Lee, T. L., Grinshpun, S. A., Martuzevicius, D., Adhikari, A., Crawford, C. M., Reponen, T., (2006). Culturability and concentration of indoor and outdoor airborne fungi in six single-family homes. *Atmospheric Environment,* 40, 2902-2910.

Lei, Y., Fang, L., Sajid, M., Akash, M. S. H., Rehman, K., Liu, Z., Shi, W., Chen, S., (2013). Estimation of urinary concentration of Aflatoxin M1 in Chinese pregnantwomen. *Journal of Food Science,* 78, T1835–T1838.

Leong, Y. H., Rosma, A., Latiff, A. A., Ahmad, N. I., (2011). Exposure assessment and risk characterization of aflatoxin B1 in Malaysia. *Mycotoxin Research*, 27, 207–214.

Londono, V. A. G., Boasso, A. C., de Paula, M. C. Z., Garcia, L. P., Scussel, V. M., Resnik, S., Pacin, A., (2013). Aflatoxin M1 survey on randomly collected milk powder commercialized in Argentina and Brazil. *Food Control*, 34(2), 752-755.

Lye, M. S., Ghazali, A. A., Mohan, J., Alwin, N. and Nair, R.C., (1995). An outbreak of acute hepatic encephalopathy due to severe aflatoxicosis in Malaysia. *American Journal of Tropical Medicine and Hygiene*, 53, 68-72.

Markaki, P. and Melissari, E., (1997). Occurrence of aflatoxin M1 in commercial pasteurized milk determined with ELISA and HPLC. *Food Additives and Contaminants,* 14, 451-456.

Martin, M. L. and Martin, H. M., (2000). Aflatoxin M1 in raw and ultra-high-temperature treated milk commercialized in Portugal. *Food Additives and Contaminants,* 17, 871-874.

Masood, M., Iqbal, S. Z., Asi, M. R., Malik, N., (2015). Natural occurrence of aflatoxins in dry fruits and edible nuts. *Food Control,* 55, 62–65.

McCoy, L., Scholl, P. F., Schleicher, R., Groopman, J. D., Powers, C., Pfeiffer, C. M., (2005). Analysis of aflatoxin B1-lysine adducts in human serum by isotope dilution liquid chromatography tandem mass spectrometry. *Rapid Communications in Mass Spectrometry*, 19, 2203-2210.

Miller, J. D., (1995). Fungi and mycotoxins in grain: implications for stored product research. *Journal of Stored Products Research,* 31(1), 1–16.

Mohammad-Hasani, F., Mirlohi, M., Mosharraf, L., Hasanzade, A., (2016). Occurrence of aflatoxins in wheat flour specified for sangak bread and its reduction through fermentation and baking. *Quality Assurance and Safety of Crops and Foods*, 1-8.

Moss, M. O., (2002). Risk assessment for aflatoxins in foodstuffs. *International Biodeterioration and Biodegradation*, 50, 137-142.

Mykkänen, H., Zhu, H., Salminen, E., Juvonen, R. O., Ling, W., Ma, J., Polychronaki, N., Kemiläinen, H., Mykkänen, O., Salminen, S., El-Nezami, H., (2005). Fecal and urinary excretion of aflatoxin B1 metabolites (AFQ1, AFM1 and AFB-N^7-guanine) in young Chinese males. *International Journal of Cancer,* 115, 879-884.

Natamba, B. K., Wang, J. S., Young, S. L., Ghosh, S., Griffiths, J. K., (2016). HIV-infected pregnant and lactating women have higher serum aflatoxin levels than HIVeuninfected women and aflatoxin levels are higher during early postpartum than during pregnancy among HIV-infected women. *FASEB Journal,* 30, 668.665-668.665.

Ngindu, A., Johnson, B. K., Kenya, P. R., Ngira, J. A., Ocheng, D. M., Nandwa, H., Omondi, T. N., Jansen, A. J., Ngare, W., Kaviti, J. N., Gatei, D., Siongok, T. A., (1982). Outbreak of acute hepatitis by aflatoxin poisoning in Kenya. *The Lancet,* 1, 1346-1348.

Nibbelink, S. K., (1986). Aflatoxicosis in food animals: a critical review. Iowa State University Veterinarian, 48, 28-31.

Nielsen, K. F., (2003). Mycotoxin production by indoor molds. *Fungal Genetics and Biology,* 39, 103-117.

Noviandi, C. T., Razzazi-Fazeli, E., Agus, A., Böhm, J., Hulan, H. W., Wedhastri, S., Maryudhani, Y. B., Sardjono, N., Leibetseder, J., (2001). Natural occurrence of aflatoxin B1 in some of Indonesian food and feed products in Yogyakarta in year 1998-1999. *Mycotoxin Research,* 17(2), 174-177.

Nyathi, C. B., Mutiro, C. F., Hasler, J. A., Chetsanga, C. J., (1987). A survey of urinary aflatoxin in Zimbabwe. *International Journal of Epidemiology*, 16, 516-519.

Okaru, A. O., Abuga, K. O., Kibwage, I. O., Hausler, T., Luy, B., Kuballa, T., Rehm, J., Lachenmeier, D. W., (2017). Aflatoxin contamination in unrecorded beers from Kenya - A health risk beyond ethanol. *Food Control,* 79, 344-348.

Paterson, R. R. M., Kelley, J., Gallagher, M., (1997). Natural occurrence of aflatoxins and Aspergillus flavus (Link) in water. *Letters in Applied Microbiology*, 25, 435-436.

Paterson, R. R. M., Lima, N., (2010). How will climate change affect mycotoxins in food? *Food Research International*, 43, 1902-1914.

Perera, F. P., (1996). Molecular epidemiology: insights into cancer susceptibility, risk assessment, and prevention. *Journal of the National Cancer Institute,* 88, 496-509.

Peterson, S. W., Ito, Y., Horn, B. W., Goto, T., (2001). Aspergillus bombycis, a new aflatoxigenic species and genetic variation in its sibling species, A. nomius. *Mycologia,* 689–703.

Piekkola, S., Turner, P. C., Abdel-Hamid, M., Ezzat, S., El-Daly, M., El-Kafrawy, S., Savchenko, E., Poussa, T., Woo, J. C. S., Mykkänen, H., El-Nezami, H., (2012). Characterisation of aflatoxin and deoxynivalenol exposure among pregnant Egyptian women. *Food Additives and Contaminants: Part A*, 29, 962–971.

Pier, A. C., (1992). Major biological consequences of aflatoxicosis in animal production. *Journal of Animal Science*, 70, 3964-3970.

Pier, A. C., McLoughlin, M. E., Richard, J. L., Baetz, A. and Dahfren, R. R., (1985). In utero transfer of aflatoxin and selected effects in neonatal pigs. In: J. Lacey, (ed.). *Trichothecenes and other mycotoxins*. New York: Wiley. P. 495-506.

Polychronaki, N., West, R. M., Turner, P. C., Amra, H., Abdel-Wahhab, M., Mykkänen, H., El-Nezami, H., (2007). A longitudinal assessment of aflatoxin M1 excretion in breastmilk of selected Egyptian mothers. *Food and Chemical Toxicology,* 45, 1210–1215.

Polychronaki, N., Wild, C. P., Mykkänen, H., Amra, H., Abdel-Wahhab, M., Sylla, A., Diallo, M., El-Nezami, H., Turner, P. C., (2008). Urinary biomarkers of aflatoxin exposure in young children from Egypt and Guinea. *Food and Chemical Toxicology,* 46, 519-526.

Qian, G. S., Ross, R. K., Yu, M. C., Yuan, J. M., Gao, Y. T., Henderson, B. E., Wogan, G. N., Groopman, J. D., (1994). A follow-up study of urinary markers of aflatoxin exposure and liver cancer risk in Shanghai, People's Republic of China. Cancer Epidemiology, Biomarkers and Prevention, 3, 3-10.

RASFF, (2017). [Online]. [Accessed 17 June 2017]. Available from World Wide Web: http://webgate.ec.europa.eu/rasff-window/portal/.

Razzazi-Fazeli, E., Noviandi, C. T., Porasuphatana, S., Agus, A. and Böhm, J., (2004). Contamination in baby food, peanut and corn products sold at retail Indonesia analysed by ELISA and HPLC. *Mycotoxin Research,* 20(2), 51-58.

Reddy, M. K., Srivinas, T., Lakshmi, K. A., (2012). A study of bioaerosols in indoor air of food godowns of Visakhapatnam, India. *Journal of Environment Research and Development,* 6(3), 446-451.

Renwick, A. G., Barlow, S., Hertz-Picciotto, I., Boobis, A. R., Dybing, E., Edler, L., Eisenbrand, G., Grieg, J. B., Kleiner, J., Lambe, J., Müller, D. J. G., Smith, M. R., Tritscher, A., Tuijtelaars, S., van den Brand, P. A., Walker, R., Kroes, R., (2003). Risk characterisation of chemicals in food and diet. *Food and Chemical Toxicology,* 41, 1211-1271.

Revankar, S. G., (2003). Clinical implications of mycotoxins and stachybotrys. *American Journal of Medical Science,* 325, 262-274.

Rogers, A. E. (1993). Nutritional modulation of aflatoxin carcinogenesis. In: D. L. Eaton and J. D. Groopman, (eds.). The toxicology of aflatoxins: human health, veterinary, and agricultural significance. London: Academic Press. P. 207-230.

Romero, A. C., Ferreira, T. R. B., Dias, C. T. S., Calori-Domingues, M. A., Gloria, E. M., (2010). Occurrence of AFM1 in urine samples of a Brazilian population and association with food consumption. *Food Control,* 21, 554-558.

Ross, R. K., Yuan, J. M., Yu, M. C., Wogan, G. N., Qian, G. S., Tu, J. T., Groopman, J., Gao, Y. T., Herderson, B. E., (1992). Urinary aflatoxin biomarkers and risk of hepatocellular carcinoma. *The Lancet,* 339, 1413-1414.

Routledge, M. N., Kimanya, M. E., Shirima, C. P., Wild, C. P., Gong, Y. Y., (2014). Quantitative correlation of aflatoxin biomarker with dietary intake of aflatoxinin Tanzanian children. *Biomarkers,* 19, 430–435.

Roy, M., Harris, J., Afreen, S., Deak, E., Gade, L., Balajee, S. A., Park, B., Chiller, T., Luby, S., (2013). Aflatoxin contamination in food commodities in Bangladesh. *Food Additives and Contaminants: Part B-Surveillance,* 6, 17–23.

Sabbioni, G., (1990). Chemical and physical properties of the major serum albumin adduct of aflatoxin B1 and their implications for the quantification in biological samples. *Chemico-Biological Interactions,* 75, 1-15.

Sabbioni, G., Skipper, P. L., Buchi, G., Tannenbaum, S. R., (1987). Isolation and characterisation of the major serum albumin adduct formed by aflatoxin B1 *in vivo* in rats. *Carcinogenesis,* 8, 819-824.

Sadeghi, N., Oveisi, M. R., Jannat, B., Hajimahmoodi, M., Bonyani, H., Jannat, F., (2009). Incidence of aflatoxin M1 in human breast milk in Tehran, Iran. *Food Control*, 20, 75-78.

Sales, A. C. and Yoshizawa, T., (2005). Updated profile of aflatoxin and *Aspergillus* section *Flavi* contamination in rice and its byproducts from the Philippines. *Food Additives and Contaminants,* 22(5), 429-436.

Scholl, P. F., Turner, P. C., Sutcliffe, A. E., Sylla, A., Diallo, M. S., Friesen, M. D., Groopman, J. D., Wild, C. P., (2006). Quantitative comparison of Aflatoxin B1 serum albumin adducts in humans by isotope dilution mass spectrometry and enzyme linked immunosorbent assay. *Cancer Epidemiology, Biomarkers and Prevention,* 15, 823-826.

Sepahvand, A., Shams-Ghahfarokhi, M., Allameh, A., Razzaghi-Abyaneh, M., (2013). Diversity and distribution patterns of airborne microfungi in indoor and outdoor hospital environments in Khorramabad, Southwest Iran. *Jundishapur Journal of Microbiology,* 6, 186-192.

Sherif, S. O., Salama, E. E., Abdel-Wahhab, M. A., (2009). Mycotoxins and child health: The need for health risk assessment. *International Journal of Hygiene and Environmental Health*, 212, 347-368.

Shirima, C. P., Kimanya, M. E., Kinabo, J. L., Routledge, M. N., Srey, C., Wild, C. P., Gong,Y. Y., (2013). Dietary exposure to aflatoxin and fumonisin among Tanzanian children as determined using biomarkers of exposure. *Molecular Nutrition and Food Research*, 57, 1874–1881.

Shuaib, F. M. B., Jolly, P. E., Ehiri, J. E., Yatich, N., Jiang, Y., Funkhouser, E., Person, S. D., Wilson, C., Ellis, W. O., Wang, J. S., Williams, J. H., (2010). Association between birth outcomes and aflatoxin B_1 biomarker blood levels in pregnant women in Kumasi, Ghana. *Tropical Medicine and International Health,* 15(2), 160-167(8).

Skipper, P. L., Hutchins, D. H., Turesky, R. J., Sabbioni, G., Tannenbaum, S. R., (1985). Carcinogen binding to serum albumin. *Proceedings of the American Association for Cancer Research,* 26, 356-361.

Stroka, J., Anklam, E., (2002). New strategies for the screening and determination of aflatoxins and the detection of aflatoxin-producing moulds in food and feed. *Trends in Analytical Chemistry,* 21, 90-95.

Sulaiman, M. R., Yee, C. F., Hamid, A., Yatim, A. M., (2007). The occurrence of aflatoxins in raw shelled peanut samples from three districts of Perak, Malaysia. *Electronic Journal of Environmental, Agricultural and Food Chemistry*, 6(5), 2045-2052.

Sun, Z., Chen, T., Thorgeirsson, S. S., Zhan, Q., Chen, J., Park, J. H., Lu, P., Hsia, C. C., Wang, N., Xu, L., Lu, L., Huang, F., Zhu, Y., Lu, J., Ni, Z., Zhang, Q., Wu, Y., Liu, G., Wu, Z., Qu, C., Gail, M. H., (2013). Dramatic reduction of liver cancer incidence in young adults: 28 year follow-up of etiological interventions in an endemic area of China. *Carcinogenesis*, 34, 1800–1805.

Szymczak, M. G., Gorny, R. L., (2010). Bacterial and fungal aerosols in air-conditioned office buildings in Warsaw, Poland-The winter season. *Inter Journal of Occupational Safety and Ergonomics,* 16(4), 465-476.

Tannenbaum, S. R., Skipper, P. L., (1984). Biological aspects to the evaluation of risk: dosimetry of carcinogens in man. *Fundamental and Applied Toxicology*, 4, S367-S373.

Tassaneeyakul, W., Razzazi-Fazeli, E., Porasuphatana, S. and Bohm, J., (2004). Contamination of aflatoxins in herbal medicinal products in Thailand. *Mycopathogia*, 158, 239-244.

Tirado, M. C., Clarke, R., Jaykus, L. A., McQuatters-Gollop, A., Frank, J. M., (2010). Climate change and food safety: A review. *Food Research International,* 43, 1745-1765.

Tsuboi, S., Nakagawa, T., Tomita, M., Seo, T., Ono, H., Kawamura, K., Iwamura, N., (1984). Detection of aflatoxin B1 in serum samples of male Japanese subjects by radioimmunoassay and high-performance liquid chromatography. *Cancer Research,* 44, 1231-1234.

Turconi, G., Guarcello, M., Livieri, C., Comizzoli, S., Maccarini, L., Castellazzi, A. M., Pietri, A., Piva, G., Roggi, C., (2004). Evaluation of xenobiotics in human milk and ingestion by the newborn – an epidemiological survey in Lombardy (Northern Italy). *European Journal of Nutrition,* 43(4), 191–197.

Turner, P. C., (2013). The molecular epidemiology of chronic aflatoxin driven impaired child growth. *Scientifica,* Article ID 152879, 21 pages http://dx.doi.org/10.1155/2013/152879.

Turner, P. C., Dingley, K. H., Coxhead, J., Russell, S., Garner, C. R., (1998). Detectable levels of serum aflatoxin B1-albumin adducts in the United Kingdom population: implications for aflatoxin B1 exposure in the United Kingdom. *Cancer Epidemiology, Biomarkers and Prevention,* 7, 441-447.

Turner, P. C., Flannery, B., Isitt, C., Ali, M., Pestka, J., (2012a). The role of biomarkers in evaluating human health concerns from fungal contaminants in food. *Nutrition Research Reviews,* 25(1), 162–179.

Turner, P. C., Moore, S. E., Hall, A. J., Prentice, A. M., Wild, C. P., (2003) Modification of immune function through exposure to dietary aflatoxin in Gambian children. *Environmental Health Perspectives,* 111(2), 217-220.

Turner, P. C., van der Westhuizen, L., Da Costa, A. N., (2012b). Biomarkers of exposure: mycotoxins—aflatoxin, deoxynivalenol and fumonisins. In: *Biomarkers and Human Biomonitoring*, L. E. Knudsen and D. F. Merlo (eds.). Royal Society of Chemistry, Cambridge, UK.

Turner, P. C., Sylla, A.,Gong, Y. Y., Diallo, M. S., Sutcliffe, A. E., Hall, A. J., Wild, C. P., (2005). Reduction in exposure to carcinogenic aflatoxins by postharvest intervention measures in West Africa: a community-based intervention study. *The Lancet,* 365(9475), 1950–1956.

Varga, J., Kerepes, L., Kocsubé, S., Szigeti, G., Baranyi, N., Vágvölgyi, C., (2013). Examination of the diversity of indoor molds in a Hungarian student hostel. *Acta Biologica Szegediensis,* 57, 21-24.

Varga, J., Kocsube, S., Szigeti, G., Baranyi, N., Vagvoigyi, C., Despot, D. J., Magyar, D., Meijer, M., Samson, R. A., Klaric, M. S., (2014). Occurance of black Arpergilli in indoor environment of six countries. *Archives of Industrial Hygiene and Toxicology*, 65, 219-223.

Walton, M., Egner, P., Scholl, P. F., Walker, J., Kensler, T. J., Groopman, J. D., (2001). Liquid chromatography electrospray mass spectrometry of urinary aflatoxin biomarkers: characterization and application to dosimetry and chemoprevention in rats. *Chemical Research in Toxicology,* 14(7), 919-926.

Wang, J. S., Abubaker, S., He, X., Sun, G., Strickland, P. T., Groopman, J. D., (2001). Development of aflatoxin B1-lysine adduct monoclonal antibody for human exposure studies. *Applied and Environmental Microbiology,* 67(6), 2712-2717.

Wang, J. S., Qian, G. S., Zarba, A., He, X., Zhu, Y. R., Zhang, B. C., Jacobson, L., Gange, S. J., Muñoz, A., Kensler, T. W. and Groopman, J. D., (1996). Temporal patterns of aflatoxin-albumin adducts in hepatitis B surface antigen-positive and antigen-negative residents of Daxin, Qidong County, People's Republic of China. *Cancer Epidemiology, Biomarkers and Prevention,* 5, 253-261.

Warth, B., Sulyok, M., Krska, R., (2013). LC–MS/MS-based multibiomarker approaches for the assessment of human exposure to mycotoxins. *Analytical and Bioanalytical Chemistry,* 405, 5687–5695.

Watson, S., Chen, G., Sylla, A., Routledge, M. N., Gong, Y. Y., (2015). Dietary exposure to aflatoxin and micronutrient status among young children from Guinea. *Molecular Nutrition and Food Research,* 60 (3), 511-518.

WHO, (1999). Basic Food Safety for Health Workers [Online]. [Accessed 22 May 2017]. p. 25. Available from World Wide Web: http://apps.who.int/iris/bitstream/10665/65992/1/WHO_SDE_PHE_FOS_99.1.pdf?ua=1.

Wild, C. P., Gong, Y. Y., (2010). Mycotoxins and human disease: a largely ignored global health issue. *Carcinogenesis,* 31(1), 71-82.

Wild, C. P., Hall, A. J., (1998). Hepatitis B virus and liver cancer – unanswered questions. *Cancer Surveys: Infections and Human Cancer,* 33, 35-54.

Wild, C. P., Hasegawa, R., Barraud, L., Chutimataewin, S., Chapot, B., Ito, N., Montesano, R., (1996). Aflatoxin-albumin adducts: a basis for comparative carcinogenesis between animals and man. *Cancer Epidemiology, Biomarkers and Prevention*, 5, 179-189.

Wild, C. P., Hudson, G. J., Sabbioni, G., Chapot, B., Hall, A. J., Wogan, G. N., Whittle, H., Montesano, R. and Groopman, J. D., (1992). Dietary intake of aflatoxins and the level of albumin-bound aflatoxin in peripheral blood in The Gambia, West Africa. *Cancer Epidemiology, Biomarkers and Prevention,* 1, 229-234.

Wild, C. P., Jansen, L. M., Cova, L., Montesano, R., (1993). Molecular dosimetry of aflatoxin exposure: contribution to understanding the multifactorial aetiopathogenesis of primary hepatocellular carcinoma (PHC) with particular reference to hepatitis B virus. *Environmental Health Perspectives,* 99, 115-122.

Wild, C. P., Jiang, Y. Z., Allen, S. J., Jansen, L. A. M., Hall, A. J., Montesano, R., (1990a). Aflatoxin-albumin adducts in human sera from different regions of the world. *Carcinogenesis,* 11(12), 2271-2274.

Wild, C. P., Jiang, Y. Z., Sabbioni, G., Chapot, B., Montesano, R., (1990b). Evaluation of methods for quantitation of aflatoxin albumin adducts and their application to human exposure assessment. *Cancer Research,* 50, 245-251.

Wild, C. P., Turner, P. C., (2002). The toxicology of aflatoxins as a basis for public health decisions. *Mutagenesis,* 17(6), 471-481.

Wild, C. P., Umbenhauer, D., Chapot, B., Montesano, R., (1986). Monitoring of individual human exposure to aflatoxins (AF) and N-nitrosamines (NNO) by immunoassays. *Journal of Cellular Biochemistry,* 30, 171-179.

Wild, C. P., Yin, F., Turner, P. C., Chemin, I., Chapot, B., Mendy, M., Whittle, H., Kirk, G. D., Hall, A. J., (2000). Environmental and genetic determinants of aflatoxin-albumin adducts in the Gambia. *International Journal of Cancer,* 86, 1-7.

Williams, J. H., Phillips, T. D., Jolly, P. E., Stiles, J. K., Jolly, C. M. and Aggarwal, D., (2004). Human aflatoxicosis in developing countries: a review of toxicology, exposure, potential health consequences, and interventions. *The American Journal of Clinical Nutrition,* 80, 1106-1122.

Wu, F., Guclu, H., (2012). Aflatoxin regulations in a network of globalmaize trade. *PLoS One* 7, e45151.

Wu, F., Khlangwiset, P., (2010). Health economic impacts and cost-effectiveness of aflatoxin-reduction strategies in Africa: Case studies in biocontrol and postharvest interventions. *Food Additives and Contaminants: Part A,* 27(4), 496-509.

Xu, L., Qian, G., Tang, L., Su, J., Wang, J. S., (2010). Genetic variations of hepatitis B virus and serum aflatoxin-lysine adduct on high risk of hepatocellular carcinoma in Southern Guangxi, China. *Journal of Hepatology,* 53, 671-676.

Yard, E. E., Daniel, J. H., Lewis, L. S., Rybak, M. E., Paliakov, E. M., Kim, A. A., Montgomery, J. M., Bunnell, R., Abudo, M. U., Akhwale, W., Breiman, R. F., Sharif, S. K., (2013). Human aflatoxin exposure in Kenya, 2007: a cross-sectional study. *Food Additives and Contaminants: Part A,* 30, 1322–1331.

Yourtee, D. M., Kirk-Yourtee, C. L., (1989). Aflatoxin Detoxification in Humans Laboratory, Field and Clinical Perspectives. *Journal of Toxicology - Toxin Review,* 8(1-2), 3-18.

Zain, M. E., (2011). Impact of mycotoxins on humans and animals. *Journal of Saudi Chemical Society,* 15(2), 129–144.

Zhu, J. Q., Zhang, L. S., Hu, X., Xiao, Y., Chen, J. S., Xu, Y. C., Fremy, J., Chu, F. S., (1987). Correlation of dietary aflatoxin B1 levels with excretion of aflatoxin M1 in human urine. *Cancer Research*, 47, 1848-1852.

In: Focus on Aflatoxins Research
Editors: P. Jan and S. Sylvester
ISBN: 978-1-53612-569-6

Chapter 2

AFLATOXINS: EXPOSURE, DETECTION AND CONTROL METHODS

Mufida Aljicevic[1,*] and Damir Ljuboja[2]
[1]Department of Medicine, University of Sarajevo, Sarajevo, Bosnia and Herzegovina
[2]Department of Medicine, Harvard Medical School, Boston, Massachusetts, US

ABSTRACT

Molds are microscopic fungi that produce metabolites and release them to the surface where growth occurs. Mycotoxins are such metabolites, with over 400 different toxins produced by approximately 350 speices of fungi. Aflatoxins are mycotoxins produced by *Aspergillus flavus* under favorable temperature and humidity conditions. Aflatoxins belong to a group of extremely toxic mycotoxins, and are therefore categorized in as Group I carcinogens by the International Agency for Research on Cancer (IARC). There are four main types of aflatoxins: B1, B2, G1 and G2, as well as two additional metabolic products: M1 and M2. The IARC has placed aflatoxin B1 on the list of human carcinogens.

* Corresponding author: mufida.aljicevic@mf.unsa.ba.

Man is exposed to the effects of aflatoxins via consumption of food that is contaminated by products of developed fungi.

Aflatoxin B1 is the strongest natural carcinogen for the liver and aflatoxin B1 and hepatitis B (HBV) are considered as major risk factors in the etiology of hepatocellular carcinoma (HCC).

The aim of this work was to determine if foods of plant origin present in our market have quantifiable concentrations of aflatoxin B1. A secondary objective was to examine whether the urine of healthy participants contains measurable aflatoxin B1, and compare this control to a population of patients with HBV and HCC.

Laboratory testing can detect the presence of aflatoxins in samples of food, as well as in the urine of patients. In our study of aflatoxin B1, samples were divided into three groups for analysis. The first group consisted of 32 samples foods of plant origin that were available in the marketplace, the second experimental group consisted of the urine of 30 patients with HCC and HBV, and the third (control) group consisted of the urine 30 "healthy" patients. All samples were processed using the Aflatoxin B1 Flow Through Rapid Test kit by Euro-Diagnostica. All samples were tested by Sensitive Aflatoxin B1 enzyme immunoassay (EIA) for screening and quantitative analysis. From a total of 32 food samples tested, first screening analysis with Flow Through Rapid Test yielded only 4 positive results, while EIA detected 11 positive results with quantitative values of aflatoxin B1 ranging from 0.23 to 11.67 ppb. In the healthy control patients and the experimental group, the values of aflatoxin B1 in the urine samples were below the detectable level of screening test (2 ppb) via the Flow Through Rapid Test. However, with Sensitive EIA testing we obtained values of 0.05 to 0.15 ppb of aflatoxin B1 in the control urine samples, and 0.05 to 0.26 ppb in the experimental group. Comparing the results clearly shows that the Aflatoxin B1 Sensitive EIA is more sensitive than Aflatoxin B1 Flow Through Rapid Test. The Rapid test is a simple and quick screening method, but is not sufficiently sensitive for quantifying aflatoxin B1 levels.

Of note, all of tested urine samples of both groups (experimental and control) showed the presence of aflatoxins B1. However, the concentration of aflatoxins was lower in the control group (maximum value of 0.15 ppb), relative to the experimental group (maximum value of 0.26 ppb).

These results suggest that aflatoxin-contaminated foods readily available in the marketplace serve as a vector for the introduction of small quantities of aflatoxins into hosts. These aflatoxins may pose a cumulative threat to human health over time given their greater concentration in patients with HBV and HCC. Thus, further research into aflatoxins and their role in the pathophysiology of HCC is indicated.

Keywords: aflatoxins, mycotoxins, sensitive EIA test, flow through rapid test

INTRODUCTION

It is a widely accepted fact that most of the foods we consume contain some potentially harmful substances. These substances can be natural ingredients inherent to food, substances that are deliberately added to foods in the production and preservation process or the result of contamination. Human illness resulting from the intake of contaminated food is one of the biggest problems faced by contemporary societies. In order to prevent food contamination, food safety programs that focus on quality control are vital. In order to minimize contamination, it is important to carefully standardize every step related to production, storage, distribution and preparation of food [1, 2].

However, despite the best control measures, there is no food that does not contain at least one harmful substance. It is therefore necessary to conduct risk assessment analysis and to determine the maximum concentration of these substances present in food in order to better understand the threat they pose.

Risk assessment is vital to the determination of the Acceptable Daily Intake (ADI) and Tolerable Daily Intake (TDI) for substances that exceed a certain threshold of toxicity. The term ADI is most commonly used to describe a "safe" level of potentially harmful substance intake, and TDI can be expressed on a daily or weekly basis. Both of these are determined on the basis of the highest levels of a potentially toxic substance that do not yield harmful effects. For risk assessment, it is important to evaluate ADI or TDI intake estimates and use the obtained data to assess whether the harmful substance can impact health [3].

Food contaminants are substances that are not intentionally added to food, and are the result of food contamination from environmental factors that come to their production, packaging, transport and processing. Mycotoxins are included in the category of natural contaminants [1].

Mycotoxins

Mycotoxins (from the Greek *mykes* "fungus" and *toxicon* "poison") are toxic secondary metabolites of mushrooms, molds and yeast. The production of secondary metabolites depends on the type of fungi, the fungal strain and its genetic characteristics. Secondary metabolites of mushrooms are synthesized in the final phase of exponential growth due to stress-inducing conditions such as inadequate moisture, temperature, poor ventilation and the presence of aggressive agents [4]. Today, several thousands of secondary metabolites of molds have been identified that do not have biochemical significance when consumed. These metabolites primarily yield competitive survival advantage for the molds over other organisms, while only some are harmful. Depending on what the organisms do, secondary metabolites are divided into antibiotics (acting on bacteria) and mycotoxins (acting on humans and animals) [5].

There are more than 400 unique secondary mold metabolites produced by about 350 fungi, which are divided into 21 classes. They are organic compounds composed of complexes from 200 to 800 kD and are not volatile at ambient temperatures. The diseases that they cause are called mycotoxicosis.

Due to their toxicity, mycotoxins introduced into the organism exhibit carcinogenic, mutagenic, immunological, hepatotoxic, nephrotoxic, dermatopic, hematopoietic and teratogenic effects. Mycotoxins have low molecular weights and do not activate the immune response, such that the human body is unprotected from their activity. Mycotoxins are a big problem for humans and animals in many countries around the world. According to FAO (Food and Agriculture Organization) estimates, 25% of the food produced in the world is contaminated with mycotoxins [6].

The most important mycotoxins include aflatoxins, ochratoxin A, fumonisin B1, zearalenone, T-2 toxin, and patulin. Mycotoxin combinations have a stronger biological effect than individual mycotoxins [7, 8, 9, 10].

Man is the end consumer of several environmental mycotoxins. A particular public health problem is residual mycotoxins in meat, milk and animal products that can affect human health. Therefore, it is necessary to know the preventive measures that safeguard against contamination of food with mycotoxins. These preventive measures are implemented at the level of cultivation as well as harvesting and storage of grains to reduce contamination risk. Until recently, mycotoxins have been a more “hidden” health problem for humans and animals [11]. Mass production of cereals, legumes, seeds and fruits that are essential for feeding a growing population, especially in developing countries, have resulted in the increased prevalence of mycotoxicoses and brought their public health impact to the forefront of medicine [12, 13].

AFLATOXINS

Aflatoxins are metabolites of saprophytic fungi of the genus *Aspergillus*, which contaminate food in areas with hot and humid climates. These environmental conditions are optimal for the growth of fungi and toxin production. The physical qualities of aflatoxins are such: they are solid at room temperature, colorless or pale yellow, crystalline in structure, thermostable and insoluble in water [14].

Aflatoxins are one of the most toxic substances that occur in nature, and also belong to the group of the most potent human carcinogens that are associated with many diseases. They belong to a group of extremely toxic mycotoxins [15, 16, 17, 18].

Aflatoxin was isolated in 1961, and its name comes from the name of the mold Aspergillus flavus toxin. Aflatoxins are primarily produced by *Aspergillus flavus*, but some strains of *Aspergillus parasiticus* and related species *Aspergillus nomius*, *Aspergillus niger*, *Aspergillus wentii*, *Aspergillus ruber* and *Aspergillus ohraceus* [19].

In fact, members of the genus Aspergillus, which produce aflatoxins, are divided into three groups:

1. The first group makes up *Aspergillus flavus*, *Aspergillus flavus* var. *parvisclerotigenus*, *Aspergillus parasiticus*, *Aspergillus toxicarius*, *Aspergillus nomius*, *Aspergillus pseudotamarii*, *Aspergilus zhaoqingensis*, *Aspergillus bombycis* i ascomicete genus *Petromyces* (Aspergillus section *Flavi*).
2. The second group consists of *Emericella astellata* i *Emericella venezuelensis* which belonging to genus *Emericella* (Aspergillus section Nidulantes).
3. The third group includes *Aspergillus ochraceoroseus*.

The most dominant species in Africa and Asia is *Aspergillus flavus*, and in America *Aspergillus parasiticus*. Aflatoxins are also produced by *Penicillium* (*Penicillium puberulum*, *Penicillium variable, Penicillium frequentans*) and some members of genus *Rhizopus* [20].

Aspergillus spp.

Aspergillus is anamorphic (asexulant) filamentous organism that propagates through asexual spores or conidia arranged in chains. Mycelium consists of grenades and septated hyphae. At the top of the air hyphae is the conidiofor, which does not branch. A conidioform is formed of a number of short radial-shaped cylindrical brackets called filialides. The basis of the conidiofore is the parent vegetative hyphae called the foot cell (the cell of the foot). Conidiofore are similar structures to hyphae and eventually expanded to form projection vesicles [21, 22].

This species is characterized by septic hyphae from which an undivided conidiofore emerges. Conidiophores can be rough or smooth, hyaline or pigmented. One or two lines of phialides form the surface of the vesicle. Chains of round or oval, pigmented conidia are produced from the top phialides. The appearance of the conidian head can be columnar to radial. Conidia contaminates different crops by direct contact (e.g., peanuts), or indirectly by spreading by wind or insects (e.g., corn) [23, 24, 25].

In silos, some insects and mites can passively transfer spores and parts of the mycelia of the fungus on their bodies, and other insects and mites transmit aflatoxins through their slow digestive systems. Contaminated material from the field comes into silos where some fungi find favorable conditions for development (e.g., *Aspergillus*), while others fail to develop due to changed conditions. The activity of insects and mites in the silos increases the moisture content of the grains and thereby creates favorable conditions for the reproduction of fungi.

Although the basic characteristic of aspergillus is to grow rapidly, there are large variations in in their growth rate, which is an important differentiation characteristic of the species [26].

Production of Aflatoxin

The growth of fungi and aflatoxin production are influenced by various physical, chemical and biological factors. Fungi produce aflatoxins at a temperatures of 12 - 40°C, at pH 3.5 to 8.0 and with the optimal water activity of about 0.99 AWA. Nutritional factors, such as sources of carbohydrates and protein, also affect the production of aflatoxin, but the mechanisms by which these factors regulate the biosynthesis of aflatoxins are not known. Moldy food can trigger aflatoxicosis which involves the formation of toxic, immunosuppressive, mutagenic, teratogenic and carcinogenic effects [27].

Temperature Effects on Aflatoxin Production

Aspergillus flavus is a very widespread mold in nature. Spores of this mold are found in the soil, in organic matter and in grain, especially on arahids (from the Greek *arachis hypogaea* "peanuts") and corn. The mold sporulates at substrate humidity greater than 15% and at a temperature of 32°C. If the temperature is changed, or there is a lower relative humidity, the substrate is not suitable for the production of toxins [28].

It takes 48 hours under optimal conditions of moisture and temperature in a medium to produce such quantity that the substrate becomes toxic. Like other molds, *Aspergillus flavus* easily contaminates the substrate that has been damaged by the influences of drought, mechanical trauma at harvest, or insects. Contamination of molds may not be visible macroscopically. Stored wheat grain carries a substantial amount of microflora consisting of epiphytic flora of the plant itself as well as microflora that developed during storage of grain. The microflora of wheat significantly changes in the course of storage as a result of grain moisture, relative humidity of the air, temperature, and other environmental factors. In conditions of increased moisture and temperature, this microflora can develop rapidly and reduce the quality and safety of the stored material [29, 30, 31].

Climate change has a major role in the production of aflatoxins by *Aspergillus* in food. These changes affect the interaction between different mycotoxic species and toxins produced in food. Changes in ambient temperature affect the level of expression of regulatory genes (aflR and aflS) and aflatoxin production from *A. flavus* and *A. parasiticus*. A direct correlation between the expression of structural genes of early development and aflatoxin B1 was reported by Abdel-Hadi and associates in 2010. They found that temperature reacts with water activity and thereby influences the ratio of regulatory gene expression (aflR/aflS), and that this relationship is directly proportional to the production of Aflatoxin B1 [32].

Interactions between water activity and temperature have significant effects on *Aspergillus spp.* and their production of aflatoxins. Increasing the temperature up to 37°C significantly reduces the production of AFB1, regardless of the growth rate of *A. flavus* under the same conditions. By adding CO_2 with the same water activity and temperature, AFB1 production is increased. In 2016, scientist Gallo and his associates demonstrated that the production of fungal biomass and AFB1 was highest at 28°C and 0.96 aw, and that there was no fungal growth or AFB1 production at 20°C with aw values of 0.90 and 0.93. Also, AFB1 production was not detected at 37°C. Using reverse transcription quantitative PCR, the regulatory genes aflR and aflS were found to be very pronounced at 28°C, while the lowest expression was observed at 20 and 37°C, indicating that the temperature sensitivity plays a significant role in gene expression and the subsequent production of aflatoxins [33, 34].

Genetic Factors Implicated in Aflatoxin Production

Many genes and their products are involved in the production of sterigmatocystine (ST) and dihydrosterigmatocystin (DHST), which are the penultimate aflatoxin precursors. These genes are named in accordance with the substrate and product derived from nor-1 (norsolorinic acid [NOR]), norA, norB, avnA (averanti [AVN]), avfA (averufin [AVF]), ver-1 (versicolorin A [VERA], verA and verB while they are based on the enzymatic function of fas-2 (FAS alpha subunits), fas-1 (FAS beta subunits), pksA (PKS), adhA (alcohol dehydrogenase), estA (esterase), vbs (VERB synthase), dmtA (MT-i; O-methyltransferase i), omtA (O-methyltransferase A), ORDA (oxidoreductase A), cypA (P450 monooxygenase), cypX (P450 monooxygenase) and moxY (monooxygenase).

Initially, the aflatoxin regulatory gene was named afl-2 from *A. flavus* and apa-2 from *A. parasiticus*. However, it was later renamed to aflR from *A. flavus*, *A. parasiticus* and *A. nidulans* due to its role as a transcriptional activator. Previous studies have shown that aflA (fas-2), aflB (fas-1) and aflC (pksA) are responsible for the conversion of acetate to NOR. In addition, the gene umv8 is essential for the synthesis of NOR and aflatoxin production in *A. parasiticus*. The amino acid sequence of the gene is similar to the beta subunit of FAS (FAS1) from *Saccharomyces cerevisiae*. FAS creates a polyketide base during aflatoxin synthesis - hence the uvm8 gene called fas-1.

Approximately 20 aflatoxins have been isolated, but only four bear significant clinical relevance: [63] B1, B2, G1 and G2, and two additional M1 and M2 metabolic products that were isolated from milk. They were named according to the color of their fluorescence on thin-layer chromatography: Aflatoxin B was named because of a blue light with UV radiation, and aflatoxins G due to green-yellow fluorescence [35, 36]. Aflatoxins M are 4-hydroxydered aflatoxins B (found in urine, milk, blood, liver after adding aflatoxin B to food).

Aflatoxins can be defined as groups of highly oxidized heterocyclic compounds that are not equal in toxicity. Aflatoxin B1 is distinguished as the most toxic variant in human and animal food. Aflatoxins are classified as carcinogenic risk in Group 1 [37, 38, 39].

After extensive testing on many experimental animals, aflatoxins have been shown to be toxic, especially aflatoxin B1, which is confirmed as a human carcinogen. Thus, in 1993, the IARC (International Agency for Research on Cancer) put aflatoxin B1 on the list of human carcinogens. The aflatoxin carcinogenicity was the reason for a more detailed study of its importance and role in regions where there is a higher incidence of primary liver cancer (HCC). Increased levels of aflatoxins were found in peanuts and other agricultural crops in these areas [40, 41, 42, 43, 44, 45, 46].

History of Aflatoxins

Aflatoxins were isolated for the first time in 1961, and they were named after the mold itself - *Aspergilus flavus* toxin. The appearance of an unknown disease named "turkey X disease" in England in 1960 resulted in the deaths of a large number of young turkeys and other animals. On autopsy, there were signs of fulminant liver necrosis that prompted investigations to detect causative factors [18].

Physico-Chemical Properties

Aflatoxins are sensitive to alkaline and acidic solutions. They are soluble in organic solvents (alcohol, acetone, chloroform) but not water. They show stability at high temperatures, but they break down at temperatures above 250°C. By heating at 100°C in an acidic medium approximately 90% of AFB1 passes into AFB2 (a degradation product), while at 160°C only 20% of AFB1 is degraded [47].

Chemically, aflatoxins (AFTs) are difuranocoumarin derivatives in which the bifurcated group is bound to the coumarin core on one side, while the pentathon ring is affixed to AFTs and AFTs-B. Of the 20 identified AFTs, AFT-B1 and AFT-B2 were produced as A. flavus products, while AFT-G1 and AFT-G2 together with AFT-B1 and AFT-B2 were produced by A. Parasiticus. AFT-B1, AFT-B2, AFT-G1, and AFT-

G2 are the four main naturally-produced aflatoxins. AFTs-M1 and AFTs-M2 on aflatoxin B derivatives [48, 49].

AFT-B1 is resistant to heat of a wide range of temperatures, including the temperatures achieved during thermal processing of foodstuffs. The biosynthetic pathway of aflatoxins consists of 18 enzymatic steps. The participation of at least 25 genes has been successfully detected and characterized [50].

Mechanism of Action

The primary toxic effects of aflatoxins appear at the level of their interaction with genetic material. The aflatoxin molecule penetrates into the cell nucleus and interacts with DNA base pairs. It create errors in the transcription of DNA, resulting in erroneous or truncated mRNA and subsequent protein products. Metabolism plays a major role in determining the degree of toxicity. The metabolic activity of hepato-cancerous mycotoxin aflatoxin B1 results in covalent binding of aflatoxins to serum albumin [51].

Consumed aflatoxins arrive at the liver, where 70-80% is excreted in the bile ducts. After ingestion, aflatoxin is metabolized by the cytochrome p450 group of enzymes in the liver, where it is converted to many metabolic products including aflatoxicol. Reversible reductase AFB1 in soluble S-10 fractions in the liver becomes aflatoxicol. It is believed that among the metabolites of aflatoxins, aflatoxicol is the most potent agent and it is a primary contributor to the development of liver cancer. Additionally, another metabolite called aflatoxin 8.9 epoxide is also formed in the liver. The amount of this metabolite determines the susceptibility of the species, because aflatoxin 8.9 epoxide induces DNA mutation by binding to the N7 position of guanine, and results in transversion of CG to TA. Guanin-Tym transversion at the third position of the p53 gene codon (249 ser) was the most commonly modified gene in patients with hepatocellular carcinoma who consumed food contaminated by aflatoxins. The toxic effect is manifested by the binding of polar epoxide of different nucleophilic centers - proteins (Schiff bases) and genetic material, thus disrupting their normal physiological function. The reaction with proteins leads to a variety of toxic effects [52, 53, 54].

Aflatoxin B1 is a potent mutagen and is considered to be one of the most potent human mutagens. It is teratogenic to experimental animals with fruit malformations and is a cause of stillbirth.

The effect of aflatoxins on the liver can be more easily explained by the penetration of aflatoxin into the cell nucleus, which results in DNA binding and polymerase inhibition, followed by inhibition of transcription. Simply said, aflatoxins react with DNA in a way that modifies RNA production by inhibiting RNA polymerase and hermetic transcription. After this inhibition, protein synthesis is reduced. If this inhibition persists, there are significant changes in the rough endoplasmic reticulum. In the metabolism of aflatoxins, microsomal oxidase is highly implicated. It has a dual role: it acts as an AFB1 activator, and helps to convert it into active AFB2a and 2,3 epoxide AFB1 metabolites. This aflatoxin detoxification process involves the conversion of AFB1 into AFQ1, AFP1, AFM1 and other water-soluble metabolites.

Aflatoxins inhibit the function of mitochondria by inhibiting the activity of several enzymes in the Krebs cycle, and consequently lead to reduced oxygen utilization, inhibition of electron transport, oxidative phosphorylation, and inhibition of protein transport in mitochondria. Aflatoxins influence lipid metabolism in the liver by acting on the enzymes involved in their synthesis and development, which is the basis of hepatotoxicity and hepatocancerogenicity. They also affect the metabolism of vitamins, causing a reduction in vitamin A concentrations. Reduced amounts of vitamin D and B2 in the diet increase the toxicity of aflatoxins.

The body's sensitivity to aflatoxins depends largely on the detoxifying system of liver, genetics, years of exposure, and other nutritional factors.

The most important route of metabolism/detoxification of AFB1 is glutathione conjugation (GSH) catalyzed by glutathione S-transferase, resulting in AFB1-GSH conjugates that are excreted in urine or bile. Mycotoxin AFB1 exposure biomarkers are protein and DNA adducts, and serum adducts with albumin correlate with DNA adducts in the urine. The teratogenic effect of aflatoxins is due to the fact that aflatoxin has the ability to pass through the placenta and into the liver of the fetus. The toxins are metabolized by the cytochrome p450 enzyme CYP3A7. As a result, AFB1 is activated into 8,9-epoxide, which can trigger abortion as well as the increased frequency of birth defects. Aflatoxins lead to suppression of T and B lymphocytes, induce changes in cytokine synthesis and expression, suppress phagocytosis by macrophages, decrease antibody production, and reduce the host organism's resistance to infectious diseases [55].

In the organism, aflatoxins cause a weak immune response due to the small amounts of aflatoxins that are ingested with food. However, when they are consumed in large quantities (which is very rare), they cause acute aflatoxicosis, and the immune response is much more robust. This is supported by the results obtained through immunization of rabbits using aflatoxin B1 bovine serum albumin conjugate. A high titer of antibodies that express significant affinity for aflatoxin B1 is obtained from the 15th day to 4 weeks after the initial immunization. These antibodies have been used in serological methods for the detection of aflatoxins via enzyme immunoassays and immunoaffinity chromatography [56].

Many studies have shown that exposure to aflatoxins can suppress immune function, specifically the cell-mediated immune response. Studies of the immunostimulatory effects of aflatoxins on the epithelium have shown that exposure to aflatoxin decreases the activity of T and B lymphocytes, as well as reduced effector functions of macrophages and neutophils, altered synthesis of inflammatory cytokines, and reduced resistance to infectious diseases.

After the oral intake of aflatoxin B1, fecal and urinary excretion of the three aflatoxin B1 metabolites (aflatoxin M1, aflatoxin Q1 and aflatoxin B1-N7 guanine) is pronounced. They are produced as the predominant byproducts of cytochrome P450 enzyme activity [57].

Epidemics Caused by Aflatoxins

The first recorded mycotoxicosis occurred in the early 1960s in England, so called. "Turkey X disease" – a disease of turkeys (high mortality) fed with peanut flour from Brazil, which was contaminated by

Aspergillus flavus, which produced mycotoxins, later called aflatoxins. After peanuts, aflatoxins were detected in different cereals and oilseeds, including beans, cotton seeds, corn flour, and rice, as well as in livestock mixtures and in different animal species.

In 1967, a 26-person poisoning epidemic occurred in Taiwan in rural villages where the victims consumed contaminated rice for >3 weeks. They developed edema of the legs and feet, abdominal pain and vomiting. Three children between the ages of 4 and 8 died. In a retrospective analysis after the epidemic, the cause of the poisoning was discovered: those affected had consumed rice contaminated with mold. Samples of the contaminated rice were found to contain 200 ppb of aflatoxin B1.

In 1974, an epidemic of aflatoxin-induced hepatitis took place in the Gujarat and Radish states in India. It is estimated that 106 people died that year due to the effects of aflatoxins. The epidemic lasted for two months and was limited to tribal people who were primarily fed on corn; it was later confirmed that the corn contained aflatoxins at a dose of 55g/kg of body weight. The disease manifested itself with high temperature, progressive jaundice, edema, pain, vomiting and the liver necrosis. A similar epidemic caused by aflatoxins, which had repercussions on human and animal populations, was also reported in Northwestern India in 1974.

An aflatoxicosis epidemic was documented in Kenya in 1981. There were 20 hospitalized patients, 12 of whom had lethal outcomes. The daily intake of aflatoxins was in the amount of 38 mg/kg of body weight, and the length of consuming food contaminated with aflatoxins was not established [58, 59].

Since 2004, there have been more epidemics of aflatoxicosis around the world, resulting in 500 acute illnesses and 200 deaths (CDCP].

The largest number of reported cases was from the rural areas of East Kenya Province in 2004. These cases were due to the consumption of domestic maize contaminated by mold. Preliminary food testing from the affected areas have revealed the presence of aflatoxins. People consumed corn contaminated with over 4,400 ppb of aflatoxin B1, which is 220 times higher than the maximum allowed limpet [60, 61, 62].

In 2013, countries in Europe, including Romania, Serbia and Croatia, and also Bosnia and Herzegovina reported cases of milk contaminated by aflatoxins.

These cases underscore the importance of quality control measures in the food development, handling, and storage processes. The FDA does not consider it possible to achieve absolute food safety when aflatoxins are in question. This absolute certainty is defined as a maximal allowable intake of 0 ppb.

Permitted Amounts of Aflatoxins in Food

Aflatoxins can be found in food products (maize, rice) as a result of fungal contamination in hot and humid conditions. The can also be found in foods as a result of contamination during the manufacturing process, and thus long after any potential fungal contaminants were biologically active.

The European Commission has limits for the maximum levels of total aflatoxin and aflatoxin B1 in peanut, nuts, dried fruit and other products. For direct human consumption these limits are 4 μg/kg. If used as a component of foodstuffs, the limit is 10 micrograms per kilogram for total aflatoxins and 5 μg/kg for aflatoxin B1. Other countries have also established aflatoxin limits for nuts and dried fruits. These limits are 1 microgram per kilogram in Switzerland, 5 in the Netherlands and Turkey, and 20 μg/kg for the USA. These limits also require adequate methods for the detection and quantification of aflatoxins in food. When it comes to corn, the maximum tolerance limit (MTL) for aflatoxin B1 is 5.0 μg/kg in European countries.

For aflatoxin M1, the limit is 0.05 micrograms per liter of fresh milk, and 0.035 micrograms per liter for heat treated and factory milk for babies. Food for babies and small children has a allowed maximum limit of 0.01 μg/kg for aflatoxin B1 [63, 64, 65].

Epidemiological, clinical and experimental studies have found that exposure to high doses (greater than 6000 mg) of aflatoxins can cause acute toxicity with lethal effects. They have also demonstrated that exposure to small doses for a long period of time has a carcinogenic effect. There are indications of emerging acute aflatoxicosis in humans.

The acute syndrome is similar to Reye's syndrome in children, which is characterized by acute liver toxicity and encephalopathy. After the invasion of aflatoxins into the liver, lipids infiltrate hepatocytes and lead to necrosis and death of liver cells. This most likely parallels what occurs when aflatoxin metabolites react negatively with different cellular proteins in the liver, leading to the inhibition of carbohydrate and lipid metabolism and protein synthesis. In correlation with the decline in liver function, a disorder in the process of blood clotting and the synthesis of essential serum proteins in the liver occurs. Other signs of aflatoxicosis are edema of the extremities, abdominal pain, and vomiting. In the case of chronic consumption of food contaminated by aflatoxins, a high percentage of hepatocellular carcinoma has been reported.

Many studies have led to the conclusion that aflatoxins were used in chemical warfare as a "cancer bomb" in the Gulf War in 1991. This resulted is the emergence of "Gulf War Syndrome." This is demonstrated by the increased cancer morbidity of US troops who participated in the operation. However, there is no direct evidence for it [66].

Aflatoxicosis

Aflatoxins cause a variety of clinical syndromes, both in humans and in animals, and are under the umbrella term aflatoxicosis. Aflatoxicosis is a primary liver disease consisting of liver cirrhosis and induction of tumor genesis. According to an individuals degrees of exposer, they may present with:

- A peracute form of toxicosis that is occurs for several hours after consuming contaminated food and leads to collapse and death.
- Acute and subacute forms occur after 12 hours post-consumption of contaminated foods. They present with hepatotoxicosis and nephrotoxicosis. Pathology shows liver flakes with mosaic necrosis in the parenchyma with hemorrhage. Glomerulonephritis is present in the kidneys. Accidents of acute poisoning are exceptional and rare.

- The chronic form occurs after long-term consumption of contaminated food, and is characterized by hepatic necrosis, hemorrhage, and unregulated proliferation of epithelial cells. As a result, it may develop into cirrhosis or liver cancer, namely HCC.

Man is exposed to aflatoxin by consuming food that is contaminated with the products of the grown mold. Such exposure is difficult to avoid because mold growth in food is not easily prevented. Although the market supply of highly contaminated food in developing countries is low, concerns remain about possible side-effects arising from long-term exposure to low levels of aflatoxins. Cases of acute aflatoxicosis in humans have been reported from many parts of the world including Taiwan, Uganda, and India. The aflatoxicosis syndrome is characterized by vomiting, abdominal pain, pulmonary edema, convulsions, coma, cardiac and brain edema, and fatty liver. Conditions that increase the likelihood of acute aflatoxicosis in humans include the presence of aflatoxin in food, environmental conditions that favor the development of fungi in seeds, and the failure of the aflatoxin monitoring and control system [67].

In humans and animals there is an acute liver cell necrosis and bile duct hyperplasia. In mild cases, recovery is complete. In more severe aflatoxicosis, there can be rapid onset ascites and edema of the lower limbs. Sudden death can be secondary to massive bleeding in the bowels [68].

Distribution of Aflatoxicosis

It is believed that aflatoxins are one of the most common causes of hepatocellular carcinoma in tropical counties. Studies conducted in Mozambique, Sudan, the Ivory Coast, China and Korea (where contamination with aflatoxins is frequent), showed that the incidence of liver cancer is higher than that in western countries. In well-developed countries, food contamination with aflatoxins is less frequent and is usually below the threshold amount that can cause acute aflatoxicosis [69].

Aflatoxin B1, HBV, HCV and *HCC*

Currently, more than 5 billion people around the world are uncontrollably exposed to aflatoxins. Aflatoxin B1 is the most commonly produced mycotoxin in cultivation conditions. Since AFB1 is the most toxic aflatoxin, it is more studied than others [70, 71].

Three key risk factors for HCC in the high risk areas of the world are:

- chronic Hepatitis B infection (HBV)
- hepatitis C (HCV) infection
- regular consumption of aflatoxins in food

Hepatitis Bx protein (HBx protein) is a multifunctional regulatory protein that plays an important role in the HBV viral life cycle. Its expression is maintained throughout all stages of cancer, and eventually leads to chronic liver inflammation. The presence of HCV in the organism leads to chronic inflammation and can lead to cellular oncoprotein activation and inactivation of tumor suppressor gene p53. Both viruses eventually lead to chronic inflammation, which eventually leads to proliferation of hepatocytes, cirrhosis and hepatocellular carcinoma. Research has shown that HBV and aflatoxin exposure have a synergistic effect on the risk of HCC development. The risk of liver cancer in persons exposed to AFB1 and chronic infection with HBV is up to 30 times greater than the risk in individuals exposed to aflatoxin alone.

Aflatoxins can modify the pathogenicity of the HBV. Epoxidation of aflatoxin to 8, 9-epoxide by the CP450 enzyme plays a key role in the pathogenesis of hepatocellular carcinoma, especially the binding of the guanine base to the third base of the 249 p53 gene codon to form aflatoxin-N7 guanine (AFB1-N7-Gua). It was noted that aflatoxins can affect the p53 suppressor gene, which is responsible for preventing the progression of the cell cycle. It has also been noted that this type of mutation dominates in areas with a high levels of aflatoxin contamination.

The interaction between AFB1 and HBV and the subsequent risk of hepatocellular carcinoma is explained by several possible mechanisms.

HBV may induce the conversion of AFB1 to AFB1-8, 9-epoxide through a specific CP450 enzyme both directly and indirectly. The presence of HBV virus itself can promote hepatocyte sensitization during AFB1 toxicity. The aflatoxin metabolite is destructive to DNA. The direct influence of AFB1 and chronic viral hepatitis can cause DNA formations to occur. An associative occurrence of HBV infection and the appearance of serum albumin during aflatoxicosis was observed in Kenya and Gambia. However, the prevalence of the mutation in codon 249 of the p53 gene was significantly higher in patients with hepatocellular carcinoma in southwestern Nigeria. In the study, all these mutations were demonstrated in patients with hepatocellular carcinoma who were HBsAg positive [72].

Also, in regions with high exposure to AFB1, the presence of hepatitis B, mutations of p53 codon 249 have been found. Synergism between AFB1 and Hepatitis B was observed through clinical studies on HBV-transgenic mice. There was a consistent correlation between HBV infection and expression of aflatoxin metabolites. It should be emphasized that HBV has the ability to integrate into host DNA, bind to p53 protein, and propagate the vulnerability of hepatocytes and carcinogenicity AFB1 [73].

In support of the aflatoxicogenic etiology of liver cancer, the incidence of liver cancer is higher in countries where there is a high aflatoxin load in food. Therefore, many researchers suggest that it may be one of the causes of cell protooncogenesis in the human liver. Thanks to recent diagnostic methods, a high correlation was found between consuming aflatoxins and the occurrence of primary liver cancer.

Aflatoxins in Breast Milk

In the European Union there are strict restrictions when it comes to food for infants.

Maximum quantities of aflatoxin concentrations in infant formula and foods with special medical purposes are as follows: AFB1 (0.1 ng/mL), AFM1 (0,025 ng/mL), OTA (0.5 ng/mL), FB1 (200 ng/mL), patulin (10 ng /mL), ZEN (20 ng/ml), and DON (200 ng/mL) [74, 75].

Although these regulations are not based on a clearly defined risk of possible mycotoxicosis, food analysis and exposure estimates have shown that baby food in Europe can be contaminated with mycotoxins and that babies are usually the population group at most risk. It should be noted that contaminated foods, including some mycotoxins and associated aflatoxins, may be transmitted from mother to child via breast milk [76, 77].

Several studies have been conducted to determine the rate of transfer during lactation. Munoz and his colleagues observed concentrations of aflatoxin in the milk of the various stages of lactation. They have shown that the concentration of AFM1 in human milk is highest in early lactation.

If we look geographically, there are different variations of concentrations of AMF1: the lowest value is among the mothers of most European countries (low risk), while critically high concentrations are found in the Saharan region of Africa and other tropical and subtropical areas, as well as in Turkey and Egypt [78, 79].

Effects of Aflatoxin Treatment

The methods of eliminating aflatoxins can be physical or chemical. Physical removal of damaged grains, seeds or nuts reduces the amount of aflatoxins by 40-80% in the food production process. The fate of aflatoxins varies with regard to the type of thermal treatment (e.g., cooking, drying, pasteurization, sterilization and spray drying). Aflatoxins disintegrate at a temperature of 237-306°C; therefore, milk pasteurization cannot protect it from AFM1 contamination. Awasthi et al. (2012) reported that neither pasteurisation nor cooking of milk would affect the level of AFM1 in cow's milk. However, by cooking corn pulp, the amount of aflatoxins is reduced by 28%, and after boiling, the toxin level decreases by 34-53%. Pouring pistachios at 90°C, 120°C and 150°C for 30, 60 and 120 minutes reduces aflatoxin levels of about by 17-63%. If hydroxyl (0.7-1%) or bicarbonate (0.4%) is added to the tortillin, the level of aflatoxins is reduced by 50-80%. Reduction of aflatoxin content depends on the time and the combination of temperatures. Saalia and Phillips (2011) reported a reduction of 84% of aflatoxins in peanut meal when the meal is cooked in the presence of calcium chloride [80, 81].

DETECTION

Evidence of aflatoxins, even in very small quantities, can be achieved by chemical methods such as:

- Thin layer chromatography (TLC)
- Gas Chromatography (GC)
- High Pressure Liquid Column Chromatography (HPLC)
- Immunoenzymatic Method (ELISA)

The most widely used detection method worldwide is Enzyme-linked Immunosorbent Assay (ELISA). It is a simple, fast method that is economically and ecologically acceptable and has the ability to analyze a large number of samples quickly. Aflatoxins are commonly detected on the basis of their absorption and emission spectra, with peak absorption at 360 nm. B toxins show blue fluorescence at 425 nm, while the G toxins exhibit green fluorescence at 540 nm under UV light [82, 83, 84].

Thin layer chromatography (TLC) is one of the oldest techniques used to detect aflatoxins, and liquid chromatography of high performance (HPLC), liquid chromatography mass spectroscopy (LCMS), and enzyme-linked immunoassays (ELISA) methods are most commonly used to detect aflatoxins. ELISA can be used for the identification of aflatoxins based on the AFB1-lysine (AFB1 toxin metabolite) concentration in blood. Specifically, the test reveals the level of AFB1 starting at 5 pg/mg of albumin, making it an effective routine screening test. In addition, several biosensors and immunoassays are being developed to detect ultra-low traces of aflatoxins in order to ensure food safety [85, 86, 87, 88].

Kinetics of Degradation

Various methods, including both chemical and physical and biological routines, are used to effectively degrade, mitigate and manage aflatoxins.

Aflatoxins AFB1 and AFG1 can be completely removed by using ozone at 8.5-40 ppm at different temperatures, but AFB2 and AFG2 are not affected by this method. The degradation of aflatoxins follows the first order kinetics. Microbiological and enzymatic degradation is preferred for the biodegradation of aflatoxins due to its environmental impacts. Bacteria *Flavobacterium* aurantiacum removes AFM1 from milk and *Nocardia asteroides* converts AFB1 into a fluorescent, nontoxic product. *Rhodococcus species* are capable of degrading aflatoxins. Bacterial ability to degrade AFB1 occurs in the following order: *R. ruber* < *R. globularulus* < *R. coprophilus* < *R. gordoniae* < *R. pyridinivorans* and < *R. erythropolis*. Fungi such as *Pleurotus ostreatus*, *Trametes versicolor*, *Trichosporon mycotoxinivorans*, *Trichoderma strain* and *Armillariella tabescens* are known to transform AFB1 into less toxic forms. In addition, the recombinant lactase produced by *A. niger* D15-Lcc2 # 3 (118 U/L) leads to 55% AFB1 reduction within 72 hours [89].

Prevention

There are several key prevention strategies for reduce aflatoxin prevalence in food supplies. They include:

- Application of physical methods: mechanical removal, grain cleaning, high temperature application, UV rays, X rays, and microwaves.
- Use of adsorbs which can bind mycotoxins and reduce their biological activity. These include calcium-aluminosilicate and natural zeolite filosilicate (used to reduce concentration of AFB1 and ZEA), and bentonite (successful in binding AFB1).
- Chemical methods: use of acids, alkalis, oxidants, reductive agents, chlorinated substances and formalin for the purpose of degradation of mycotoxins. However due to high prices and their residues in food, these methods are rarely used.

- Biological methods include the use of absorption agents, enzymes and microorganisms, including some bacteria to metabolize aflatoxins into inactive compounds [90].

Biological Management and Control Strategies

The principle of BioControl exclusion of toxic strains of *A. flavus* involves the use of non-toxic strains to reduce the contamination of aflatoxins in corn. Using BioControl agents such as *Bacillus subtilis*, *Lactobacillus spp.*, *Pseudomonas spp.*, *Ralstonia spp.*, and *Burkholderia sp*p. increases the effectiveness of aflatoxin control and management. Several strains of *B. subtilis* and *P. solanacearum* isolated from the nerizosferic type of corn affect the elimination of aflatoxins. Thc usc of biological control methods in crops has been approved by the EPA in the US. Two commercial non-toxic *A. flavus* strains (Afla-Guard® and AF36®) are used for the prevention of aflatoxin occurrence in peanut, corn and cottonseed. Good Agricultural Practice (GAP) also helps to control toxins to a great extent, and includes timely planting, adequate nutrition of plants, weed control, crop rotation, and efficiently controlling *A. flavus* infection in the field. Biological control appears to be a promising approach to managing aflatoxin in peanuts using *Trichoderma spp.* Experimental models suggests a significant reduction of 20-90% in aflatoxin infections since these methods became available [91].

Material and Methods

In 2009-210, we conducted research regarding the prevalence of aflatoxins in foods available in the marketplace. Thirty-two food products were randomly collected from different supermarkets and were analyzed for presence AFB_1. From a total of 32 food samples, 13 food samples were originally from Bosnia and Herzegovina, and 19 samples of food were imported from other countries. All samples were tested using two methods:

the Aflatoxin B_1 Flow-Through Rapid Test and the Aflatoxin B_1 EIA sensitive test. Further processing depended on the type of analysis.

Approximately 50-100 g of sample food was ground and pulverized into a fine homogenous compound. The samples were centrifuged for 15 minutes at 2000 rpm/min. The supernatant was collected. All samples were processed by Aflatoxin B_1 Flow Through Rapid Test/comercial kit obtained from Euro-Diagnostica B.V. Netherlands. This is a competitive sequential enzyme immunoassay test for the screening of food samples for the presence of Aflatoxin B_1. Rapid test has a detection minimal limit of 2 ppb (μg/kg). All values below that gave a negative result. The membrane-based assay kit consisted of 10 devices each pre-coated with rabbit antibodies to mouse IgG. The specific antibodies were bound by immobilized anti-mouse antibodies. On addition of the sample, AFB_1 (in the sample extract) then bound to the mouse anti-AFB_1 antibodies. After the sample extract was completely adsorbed through the membrane the enzyme labeled AFB_1 bound to free mouse anti-AFB_1 antibodies. The unbound conjugate (enzyme labeled AFB1) was removed by a washing step. A chromogen substrate (tetramethylbenzidine-TMB) was then added. Bound enzyme transformed the chromogen substrate into a blue colored product, which appeared as a line. The results were visually interpreted within 5 minutes after the addition of the substrate. If the sample was negative for aflatoxin, then there were two blue colored lines (C and T). If the sample is positive it will appear only one blue colored line C (Figure 1).

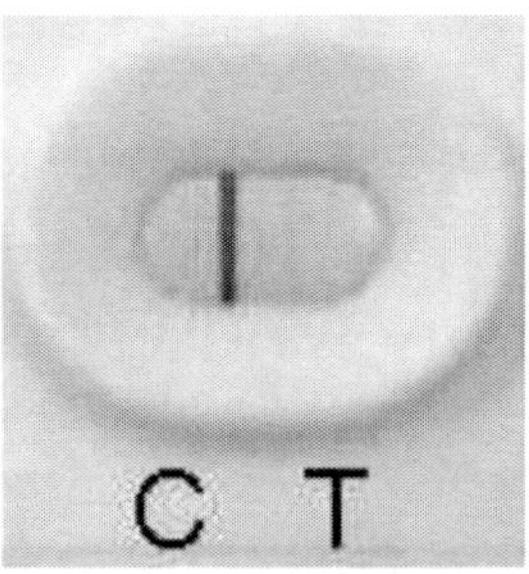

positive

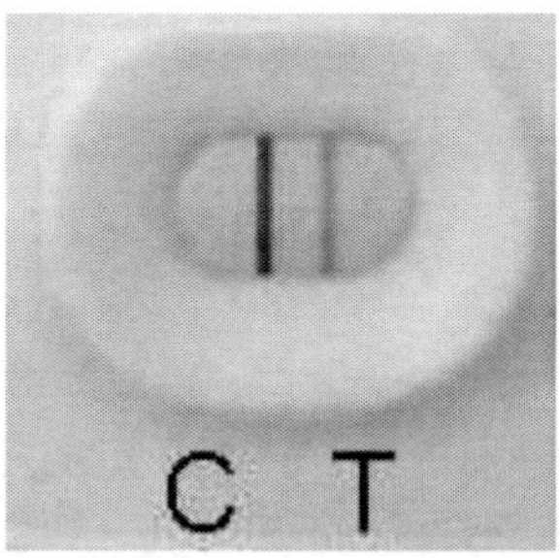

negative

Figure 1. Interpretation results of Aflatoxin B_1 Flow Through Rapid Test.

Subsequently, all samples were tested by Sensitive Aflatoxin B_1 EIA (commercial ELISA kit for AFB_1), which was obtained from Euro-Diagnostica B. V. Netherlands. This is a qualitative and much more sensitive method than the previous test. The sensitivity is reflected in what can be measured and the lowest concentration of aflatoxin B_1 in the analyzed samples. All food samples were ground to fine powder and subsamples of 20 g were mixed with 60 ml of 80% methanol and shaken in a blender at high speed for 3 minutes. The extract was allowed to settle and then filtered through filter paper. The filtrate was diluted to 1:4 concentration with dilution buffer and used for ELISA analysis. The microtiter plate based EIA kit consist of 12 strips, each 8 wells, pre-coated with specific antibodies (mouse anti-Aflatoxin B_1). Before the first incubation step, Aflatoxin B_1 standards or sample were added to the pre-coated wells. The free Aflatoxin B_1 (in the standard solution or in the sample) was bound by the specific antibody binding sites. After a washing step, and before the second incubation step, enzyme labeled aflatoxin (enzyme conjugate) was added into the wells. The enzyme conjugate was bound by free specific antibody binding sites. After the second incubation step, the non-bound (enzyme labeled) reagents were removed in a washing step. The amount of bound enzyme conjugate was visualized by the addition of chromogen substrate (tetramethylbenzidine-TMB) to each well and incubated at room temperature for 30 minutes. Bound enzyme transforms the chromogen into a colored product. The supstrate reaction is stopped by the addition of sulphuric acid. The color intensity is measured photometrically at 450 nm and is inversely proportional to the Aflatoxin B_1 concentration in the sample.

RESULTS

All 32 food samples are listed in Table 1. From a total of 32 food samples tested by first screening test (Aflatoxin B_1 Flow Through Rapid Test), only 4 samples (11%) were positive and 28 samples (89%) were negative.

Table 1. Food samples qualitatively analyzed for Aflatoxin B1

1. Flax seed, Canada w/n 9	12. Red pepper, Hungary w/n 20	23. Tea, B&H w/n 31
2. Coffee, B&H w/n 10	13. Corn flour, Slovenia w/n 21	24. Dried raisins, Iran w/n 32
3. Tea, Croatia w/n 11	14. Ground pepper, India w/n 22	25. Pistachio, Turkey w/n 33
4. Cinnamon, Turkey w/n 12	15. Nut-meg, Slovenia w/n 23	26. Coffee, B&H w/n 34
5. Wheat semolina, BH w/n 13	16. Barley porridge, B&H w/n 24	27. Popcorn, Hungary w/n 35
6. Cocoa powder, Croatia w/n 14	17. Tea, B&H w/n 25	28. Tea, Ausria w/n 36
7. Peanut, B&H w/n 15	18. Sunflower seed, Hungary w/n 26	29. Rice, Slovenia w/n 37
8. Coffee, Croatia w/n 16	19. Coffee, B&H w/n 27	30. Coffee, B&H w/n 38
9. Hazelnut, B&H w/n 17	20. Sesame, Turkey w/n 28	31. peanuts in shell, China w/n 39
10. Tea, B&H w/n 18	21. dried figs, Thailand w/n 29	32. Coffee, B&H w/n 40
11. Peanut, B&H w/n 19	22. Rice, Thailand w/n 30	* w/n – working number

Table 2. The positive food samples

1.	Food sample No. 9	/flax seed/
2.	Food sample No. 20	/red pepper/
3.	Food sample No. 23	/nut-meg/
4.	Food sample No. 24	/barley porridge/

Table 2 shows the positive food samples in which the AB1 was in a concentration of 2 ppb and higher. If the concentration of aflatoxin in food sample was less than 2 ppb, the test result was negative.

The values of Aflatoxin B_1 in the urine are much lower than 2 ppb and this rapid test in urine samples of experimental and control group was negative.

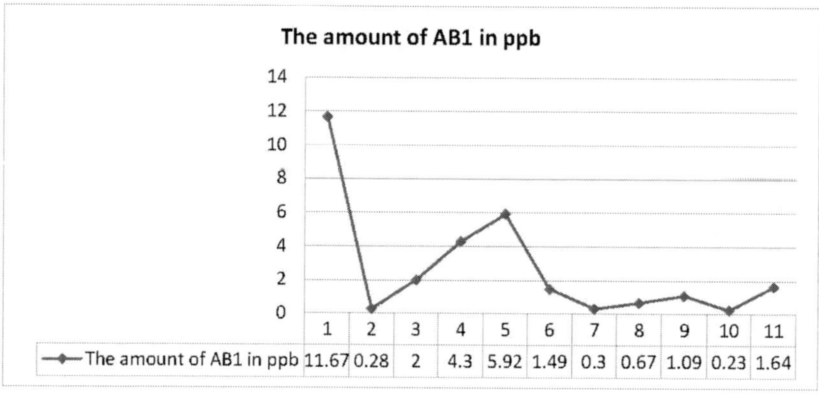

Figure 2. Positive results of food samples and the values of Aflatoxin B1.

The Sensitive Aflatoxin B_1 EIA test detected the presence of aflatoxin B_1 in 11 (34%) food samples; 21 food samples (89%) were negative. Figure 2 shows the values of AB_1 in the positive food samples.

The application of Aflatoxin B_1 Sensitive EIA test obtained different values of aflatoxin B_1 in the urine samples. Figure 3 clearly shows that in all 30 urine samples of the experimental group (30 patients with HCC and HBV infection) the recorded values of aflatoxin B_1 ranged from 0.05 ppb to 0.26 ppb.

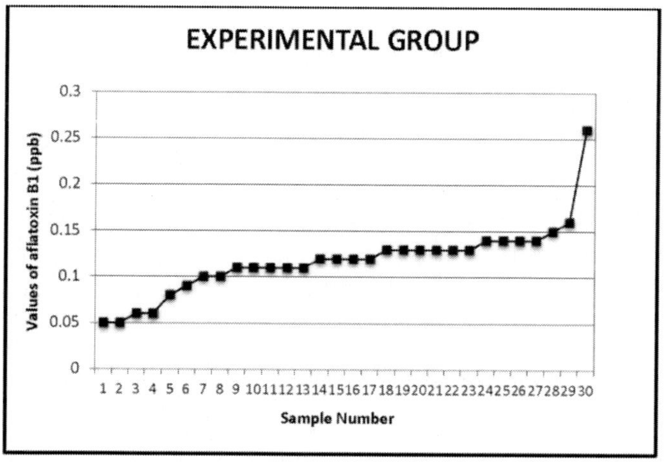

Figure 3. The concentration of Aflatoxin B1 in the urine of Experimental group.

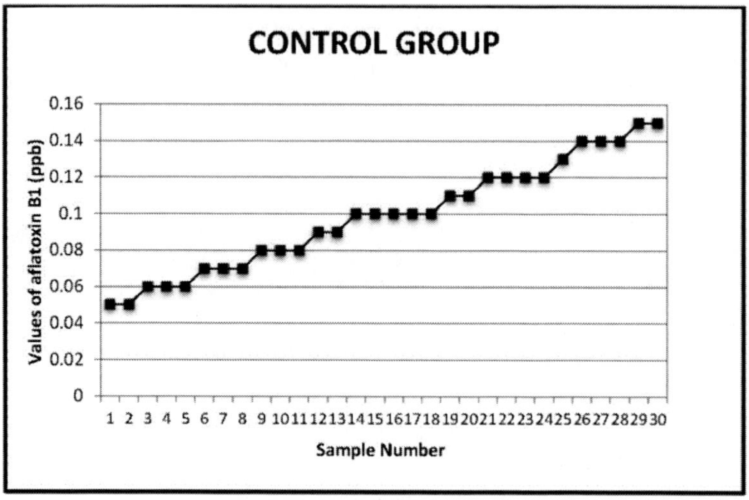

Figure 4. The concentration of Aflatoxin B1 in the urine of Control group.

Figure 4 shows the values of aflatoxin B_1 in the control group (30 healthy subjects). In that group the values of aflatoxin B_1 ranged from 0.05 ppb to 0.15 ppb.

Discussion

In the last decade, rapid immunoassay-based tests have increasingly been used in the food sector. Screening tests have proven to be very practical for use in the quality control of food via their detection of the presence of aflatoxins. The aims of this study were to determine whether there are aflatoxin B_1 in the plant food that can be found in the market, and whether there are aflatoxin B_1 in the urine of patients as a result of consuming food contaminated with aflatoxin. We compared the results, the concentrations of aflatoxin B_1 in the urine of healthy subjects with concentrations of aflatoxin B_1 in patients with hepatocellular carcinoma and hepatitis B viral infections [92, 93, 94, 95].

In our study we tested 32 food samples, 30 urine samples of patients with HCC and HBV infection, and 30 controls urine samples of healthy subjects for the presence of aflatoxin B1. From a total of 32 food samples tested by first screening test - Aflatoxin B_1 Flow Through Rapid Test, only 4 samples (11%) were positive and 28 samples (89%) were negative. The Sensitive Aflatoxin B_1 EIA test detected the presence of aflatoxin B_1 in 11 (34%) food samples and 21 (89%) food samples were negative. The values of Aflatoxin B_1 in the urine are much lower than 2 ppb (μg/kg) and this rapid test in urine samples of experimental and control groups was negative. The Sensitive Aflatoxin B_1 EIA test detected the presence of aflatoxin B_1 in all urine samples of patients with HCC i HBV infection, with values of aflatoxin B_1 from 0.05 ppb to 0.26 ppb. Also, aflatoxin B_1 was detected in urine samples of the group of healthy subjects with values ranged from 0.05 ppb to 0.15 ppb.

In India 2011, Mohammad and others were doing the similar research. All samples (food and urine) were screened for levels of aflatoxin B_1 by several methods, including the competitive enzyme linked immunosorbent assay. Also in India in 2011, Chamarty VR and others were doing a multi-center study conducted in sorghum (cereal crop) to evaluate natural contamination of aflatoxin B_1, also by deploying the method of indirect competitive ELISA [96]. In Malaysia 2011 Kasa RN and others were randomly collected food products from different supermarkets and were analyzed for presence of aflatoxin B_1 by enzyme-linked immunosorbent assay (ELISA) [97].

The general population is exposed to aflatoxins primarily by eating contaminated food.

Several ecological studies have also shown a significant association between dietary aflatoxin exposure and hepatocarcinogenesis. The results of biomarker-based cohort study from Shanghai has demonstrated a strong association between baseline levels of urinary aflatoxin B metabolites, especially the major AFB-DNA adduct, and HCC risk [98]. In Singapore 2005 Zheng et al. also analyzed food samples for the presence of aflatoxin B1 using various laboratory techniques and emphasized the advantage of rapid tests [99].

Comparing our results clearly show that the Aflatoxin B1 Sensitive EIA is far more sensitive than the Aflatoxin B1 Flow Through Rapid Test for the detection of aflatoxins. The Rapid test is a simple and very quick screening method for preliminary results, but it is not sufficiently sensitive. The Sensitive Aflatoxin B1 EIA test is superior in its ability to detect the minimum concentration of aflatoxin B1 in the analyzed samples.

This study provides useful information about the risk of aflatoxin B_1 hazards in different kinds of food products. For the improvement of the health of the entire population, it is necessary to stress the importance of the need for regular monitoring and more stringent food safety systems in order to control aflatoxins at the lowest possible levels.

From the data provided by the Food Safety Agency of Bosnia and Herzegovina, which continue to cooperate with food control laboratories in Bosnia and Herzegovina in the field of public health, veterinary medicine, agriculture and the private sector in 2015, we conclude that contamination with aflatoxins is not a significant problem in this country [100]. A total of 482 laboratory analyses were carried out to detect the presence of mycotoxins. Of these samples, 12 (2.48%) were found to not meet current regulations. Samples that were not in compliance with the applicable regulations had the content of aflatoxin M1 greater than the maximum allowable quantity. During the testing of dairy products and similar products, 203 samples were taken, excluding fat and oil products, of which 12 were classified as inadequate, representing 5.91%.

CONCLUSION

The results obtained show that the contaminated food enter a certain amount of aflatoxins B1 and if they are in small quantities pose a threat to human health and animals - aflatoxins B1 have a cumulative effect. Finding aflatoxins B1 in the urine of both groups of patients indicates that this mikotoksin is evident in the food that is in our market.

REFERENCES

[1] Turner, P. C., Flannery, B., Isitt, C., Ali, M. & Pestka, J. (2012). The role of biomarkers in evaluating human health concerns from fungal contaminants in food. *Nutr Res Rev*, *25*, 162–179.

[2] Egmond, H., Schothorst, R. & Jonker, M. (2007). Regulations relating to mycotoxins in food. *Anal Bioanal Chem*, *389*, 147-157.

[3] Knežević, Z. (2007). Food contamination with organic noxa. Hrvatski časopis za javno zdravstvo. Hrana i zdravlje. Vol 3, Broj 9. [*Croatian Journal of Public Health. Food and health.* Vol. *3*, Number 9.].

[4] Bešlagić, E. i saradnici. (2010). Medicinska mikrobiologija: Opća medicinska mikologija. Medicinski fakultet Univerziteta u Sarajevu. 441-450. [Medical microbiology: General medical mycology. Faculty of Medicine, University of Sarajevo. 441-450.].

[5] Aljicevic, M., Beslagic, E., Zvizdic, S., Hamzić, S. & Rebic, V. (2008). *Mycotoxins. HealthMed*, *2*(3), 176-182.

[6] Moss, M. O. (1998). Recent studies of mycotoxins. *Journal of Applied Microbiology Symposioum. Supplement*, *84*, 62S-76S.

[7] Wu, F., Groopman, J. D. & Pestka, J. J. (2014). Public health impacts of foodborne mycotoxins *Annu. Rev. Food Sci. Technol*, *5*, 351– 372.

[8] Mally, A. (2012). Ochratoxin and mitotic disruption: Mode of action analysis of renal tumor formation by ochratoxin A *Toxicol Sci*, *127*, 315– 330.

[9] Metzler, M., Pfeiffer, E. & Hildebrand, A. A. (2010). Zearalenone and its metabolites as endocrine disrupting chemicals. *World Mycotoxin J*, *3*, 385– 401.

[10] Mally, A., Solfrizzo, M. & Degen, G. H. (2016). Biomonitoring of the mycotoxin Zearalenone: current state-of-the art and application to human exposure assessment, *Arch Toxicol*, *90*, 1281– 1292.

[11] Reyneri A. (2006). The Role of Climatic Condition on Micotoxin Production in Cereal. *Veterinary Research Communications*, *30*(1), 87-92.

[12] Kosalec, I. & Pepeljnjak, S. (2004). Najznačajniji mikotoksini i mikotoksikoze. Praxis veterinaria, 52(3), 169-182. [The most significant mycotoxins and mycotoxicoses. *Praxis veterinaria*, *52* (3), 169-182.].

[13] Surai, P. F. & Mezes, M. (2005). Mycotoxins and immunity: Theoretical consideration and practical applications. *Praxis veterinaria*, *53*(1-2), 71-88.

[14] Eastham, S., Ingham, S., Hallett, R., et al. (2008). Towards aflatoxins: a formal synthesis of aflatoxin B1. *Tetrahedron*, *64*(5), 936-948.

[15] Turner, N. W., Subrahmanyam, S. & Piletsky, S. A. (2009). Analytical methods for determination of mycotoxins: a rewiew. *Analytica Chimica Acta*, *26* (632/2), 168-80.

[16] Kerry Scott Lane. (1999). Aflatoxin, Tobacco, Ammonia and the p53 Tumor-Suppressor Gene: Cancer's Missing Link? *Medscape General Medicine*, *1*(2), 1-3.

[17] Bennett, J. W. & Klich, M. (2003). Mycotoxins. *Clin Microbiol Rev*, *16*, 497–516.

[18] Bennett, J. W., Kale, S. & Juijiang, Y. (2015). Aflatoxins: background, Toxycology, and Molecular Biology.

[19] Vidal, A., Marin, S., Ramos, A. J., Cano-Sancho, G. & Sanchis, V. (2013). Determination of aflatoxins, deoxynivalenol, ochratoxin A and zearalenone in wheat and oat based bran supplements sold in the Spanish market. *Food and Chemical Toxicology*, *53*, 133–138.

[20] Frisvad, J. C., Skouboe, P. & Samson, R. A. (2005). Taxonomic comparison of three different groups of aflatoxin producers and a new efficient producer of aflatoxin B1, sterigmatocystin and 3-O-methylsterigmatocystin, Aspergillus rambellii sp. Nov. *Systematic and Applied Microbiology*, *28*, 442-453.

[21] Bryden, W. L. (2012). Mycotoxin contamination of the feed supply chain: Implications fr animal productivity and feed security. *Animal Feed Science and Technology*, 134-158.

[22] Razzaghi-Abyaneh, M., et al. (2006). A survey on distribution of Aspergillus section Flavi in corn field soils in Iran: Population patterns based on aflatoxins, cyclopiazonic acid and sclerotia production. *Mycopathologia*, *161*, 183-192.

[23] Miksits, K. & Hahn, H. (2003). Basiswissen Medizinische Mikrobiologie und Infektologie. 3. Auflage Berlin, 221-223.

[24] Kayser, F. H., Bienz, K. A., Eckert, J. & Zinkernagel, R. M. (2005). Fungi as Human Pathogens. Aspergillus (Aspergillosis). *Medical Microbiology Thieme*, 364-366.

[25] Kayser, F. H., Bienz, K. A., Eckert, J. & Zinkernagel, R. M. (2005). Fungi as Human Pathogens. Asperillus (Aspergillosis). *Medical Microbiology Thieme*, 364-366.

[26] Uzunovic-Kamberović, S. (2009). Medicinska mikrobiologija: Aspergiloza. Fojnica; 676-684. [Medical Microbiology: Aspergillosis. Fojnica; 676-684.].

[27] Eastham, S., Ingham, S., Hallett, R., et al. (2008). Towards aflatoxins: a formal synthesis of aflatoxin B1. *Tetrahedron*, *64*(5), 936-948.

[28] Kalinić, S. i suradnici. (2013). Medicinska mikrobiologija, Aspergiloza. Medicinska naklada Zagreb; 561-563. [Kalinic S. and associates. (2013) Medical Microbiology: Aspergillosis. Medicinska naklada Zagreb; 561-563.].

[29] Carroll, K., Morse, S., Mietzner, T. & Miller S. (2016). Jawetz, Melnick & Adelberg's medical microbiology: Aspergillosis. 27 Edition, Lange, 690-691.

[30] Hedayati, M. T., Pasqualatto, A. C., Warn, P. A., et al. (2007). Aspergillus flavus: human pathogen, allergen and mycotoxin producer. *Microbiology*, *153*, 1677-1692.

[31] Cypionka, H. (2002). Grundlagen den Mikrobiologie Berlin – Heilderberg – New York. 264-265.

[32] Halt, M., Šutić, M. Zastupljenost gljiva koje roizvode aflatoksin na žitaricama i nekim njihovim prerađevinama Slavonsko – Baranjskog područja. Simpozijum o mikotoksinima, Sarajevo 14.12.1979. ANU Bosne i Hercegovine, Posebna izdanja, Knjiga 10. Odjeljenje medicinskih nauka. [Representation of fungi that produce aflatoxin on cereals and some of their products from the Slavonian-Baranja region. Symposium on Mycotoxins, Sarajevo 14.12.1979. ANU Bosnia and Herzegovina, Special Edition, Book 10. Department of Medical Sciences.].

[33] Abdel-Hadi, A1., Carter, D. & Magan, N. (2010). Temporal monitoring of the nor-1 (aflD) gene of Aspergillus flavus in relation to aflatoxin B_1 production during storage of peanuts under different water activity levels. *J Appl Microbiol*, *109*(6), 1914-22.

[34] Gallo, A., Solfrizzo, M., Epifani, F., Panzarini, G. & Perrone, G. (2016). Effect of temperature and water activity on gene expression and aflatoxin biosynthesis in Aspergillus flavus on almond medium. *Int J Food Microbiol*, *217*, 162–169.

[35] Bacaloni, A., Cavaliere, Ch., Cucci, F., Foglia, P., Samperi, R., & Lagana, A. (2008). Determination of aflatoksins in hazelnuts by various sample preparation methodds and liquid chromatography-tandem mass sprectrometry. *Journal of Chromatography A*, 182-189.

[36] Fu, z., Huang, X. & Min, Sh. (2008). Rapid deterination of aflatoksins in corn and peanuts. *Journal of Chromatography A*, *1209*, 271-274.

[37] Molyneux, R. J., Mahoney, N., Kim, J. H. & Bruce, C. (2007). Mycotoxins in edible tree nuts. *Campbell International Journal of Food Microbiology*, *119*, 72–78.

[38] Ratnavathi, C. V., Komala, V. V., Kumar, B. S. V., Das, I. K. & Vishnu Patil, J. V. (2012). Natural occurrence of aflatoxin B1in sorghum grown in different geographical regions of India. *J Sci Food Agric*, *92*, 2416–2420.

[39] Krout-Greenberg, N. D., Puschner, B., Davidson, M. G. & DePeters, E. J. (2012). Preliminary study to assess mycotoxin concentrations in whole corn in the California feed supply. *J Dairy Sci*, *20136*, 1–8.

[40] Agriopoulou, S., Koliadima, A., Karaiskakis, G., Kapolos, J. (2016). Kinetic study of aflatoxins' degradation in the presence of ozone. *Food Control*, *61*, 221–226.

[41] Alberts, J. F., Gelderblom, W. C. A., Botha, A. & van Zyl, W. H. (2009). Degradation of aflatoxin B1 by fungal laccase enzymes. *Int J Food Microbiol*, *135*, 47–52.

[42] Peraica, M. & Plaština, R. (2000). Exposure to mycotoxins and malignant tumors. *Arh hig rada toksikol*, *51*, 131-139.

[43] Iqbal, S. Z., Mustafa, H. G., Asi, M. R. & Jinap, S. (2014). Variation in vitamin E level and aflatoxins contamination in different rice varieties. *J Cereal Sci*, *60*, 352–355.

[44] Bennett, J. W. & Klich, M. (2003). Mycotoxins. *Clin Microbiol Rev*, *16*, 497–516.

[45] Ferlay, J., Shin, H. R., Bray, F., Forman, D., Mathers, C. & Parkin, D. M. (2010). Estimates of worldwide burden of cancer in 2008, GLOBOCAN 2008. *Int J Cancer*, *127*(12), 2893-917.

[46] Wu, H. C. & Santella, R. (2012). The Role of Aflatoxins in Hepatocellular Carcinoma. *Hepat Mon.*, *12*(10 HCC), e7238.

[47] Perši, N., Pledin, J., Vulić, A. & Mitak, M. (2011). Mycotoxins in cereals and food of animal origins. *Vet Str*, *42* (4), 335-345.

[48] Bennett, J. W. & Klich M. (2003). Mycotoxins. *Clin Microbiol Rev*, *16*, 497–516.

[49] Wolf-Hall, C. (2010). "Fungal and mushroom toxins," in Pathogens and Toxins in Foods: Challenges and Interventions, eds V. K. Juneja and J. N. Sofos (Washington, DC: ASM Press). 275–285.

[50] Brown, D. W., Adams, T. H. & Keller, N. P. (1996). Aspergillus has distinct fatty acid synthases for primary and secondary metabolism. *Proc Natl Acad Sci USA.*, *19*, 14873–14877.

[51] Barrett, J. (2005). Liver Cancer and Aflatoxin. New Information from the Kenuyan Outbreak. *Enviromental Health Perspectives*, *113*(12), 837-838.

[52] Bren, U., Guengerish, P. & Mavri, J. (2007). Guanine Alkylation by the Potent Carcinogen Aflatoxin B1: Quantum Chemical calculations. *Chem Res Toxicol*, *20*(8), 1134-1140.

[53] Wild, C. P. (2002). Turner PC. The toxicology of aflatoxins as a basis for public health decisions. *Mutagenesis*, *17*(6), 471-481.

[54] Essigmann, J. M., et al. (1997). Structural identification of the major DNA adduct formed by aflatoxin B1 *in vitro*. *Proc Natl Acad Sci*, *74*(5), 1870-1874.

[55] Smela, M., Hamm, et al. (2002). The aflatoxin B1 from amidopyrimidine adduct plays a major role in causing the types of mutations observed in human hepatocellular carcinoma. *Proceedings of the National Academy of Science of United States of America (PNAS)*, *99* (10), 6655-6660.

[56] Gathumbi, J. K., Usleber, E. & Martlbauer, E. (2000). Production of ultrasensitive antibodies against aflatoxin B1. *Lett Appl Microbiol*, *32*(5), 349-51.

[57] Mykkanen, H., Zhu, H., Salminen, E., at all. (2005). Fecal and urinary excretion of aflatoxin B1 metabolites (AFQ1, AFM1 and AFB-N7-guanine) in young Chinese males. *Int J Cancer.*, *115*(6), 879-84.

[58] Subhkij, Angsubhakorn: Mycotoxins and human health risks an overview. Available at: http://www.fao.org/docrep/x5036e/ x5036E05, htm Retrieved 02. 10. 2007.

[59] Aflatoxins, historical overview. Available at: http://www.itg.be/ itg/48Micotoxins, htm Retrieved 12.05.2006.

[60] Azziz-Baumgartner, E., Lindblade, K., et al. (2005). Case-Control Study of an Acute Aflatoxicosis Outbreak, Kenia, 2004. *Environmental Health*, *113*(12), 1779-1783.

[61] Outbreak of Aflatoxin Poisoning---Eastern and Central Provinces, Kenya, January-July 2004. MMWR Weekly September 3, 2004, 53(34), 790-793.

[62] Strosnider, H., Azziz-Baumgartner, E., et al. Workgroup Report: Public Health Strategies for Reducing Aflatoxin Exposure in Developing Countries. Environmental Health Perspectives December 2006, 114(12), 1898-1903.

[63] Molyneux, R. J., Mahoney, N., Kim, J. H. & Bruce, C. (2007). Mycotoxins in edible tree nuts. *Campbell International Journal of Food Microbiology*, *119*, 72–78.

[64] Ratnavathi, C. V., Komala, V. V., Kumar, B. S. V., Das, I. K. & Vishnu Patil, J. V. (2012). Natural occurrence of aflatoxin B1in sorghum grown in different geographical regions of India. *J Sci Food Agric*, *92*, 2416–2420.

[65] Krout-Greenberg, N. D., Puschner, B., Davidson, M. G. & DePeters, E. J. (2013). Preliminary study to assess mycotoxin concentrations in whole corn in the California feed supply. *J Dairy Sci*, *96*, 1–8.

[66] Stanojevič, M., Biološko, oružje., Aflatoksin. SKP – Slovenski kemijski portal (2007). [Biological Weapons: Aflatoxin. SKP - Slovenian Chemical Portal 2007.].

[67] Aflatoxins in your food – and their Effect on Your Health. Available at: htto://www.ehso.com/ehshome/aflatoxins.php retrieved 18.11.2006.

[68] Peraica, M. & Rašić, D. (2012). Acute and chronic humamn mycotoxicoses. *Krmiva*, *54* (39), 81-87.

[69] Aljičević, M. (2010). Aflatoksini u hrani biljnog porijekla kao mogući uzrok nastanka hepatocelularnog karcinoma jetre. Doktorska disertacija. Sarajcvo, Medicinski fakultet Univerziteta u Sarajevu [Aflatoxins in foods of plant origin as a possible cause of hepatocellular liver cancer. Doctoral thesis. Sarajevo, Faculty of Medicine, University of Sarajevo].

[70] Kucukcan, B., Hayrulai-Musliu, Z. Role of Dietary Aflatoxin B1 Exposure and Hepatitis B Infection on Risk of Hepatocellular carcinoma. 25.03.2015

[71] Thigpen, K. G. (2004). Hepatitis B Virus Mutation Predicts Liver Cancer. *Environmental Health Perspectives*, *112* (11), 619-620.

[72] Peraica, M. & Plaština, R. (2000). Exposure to mycotoxins and malignant tumors. *Arh hig rada toksikol*, *51*, 131-139.

[73] Iqbal, S. Z., Mustafa, H. G., Asi, M. R. & Jinap, S. (2014).. Variation in vitamin E level and aflatoxins contamination in different rice varieties. *J Cereal Sci*, *60*, 352–355.

[74] European Commission (2006). Commission Regulation (EC) No. 1881/2006 of 19 December 2006 Setting Maximum Levels of Certain Contaminants in Foodstuffs (Consolidated Version 2010-07-01), in *Off. J. EUR. Union*, L36, 5-24.

[75] SCOOP (2003). Collection of Occurence Data of Fusarium Toxins in Food and Assessment of Dietary Intake by the Population of EU Member States, Scientific Cooperation on Questions Relating to Food.

[76] Cano-Sancho, G., Ramos, A. J., Marín, S. & Sanchis V. (2012). Occurrence of fumonisins in Catalonia (Spain) and an exposure assessment of specific population groups Food Addit Contam, Part A, 29, 799– 808.

[77] Assunção, R., Vasco, E., Nunes, B., Loureiro, S., Martins, C. & Alvito, P. (2015). Single-compound and cumulative risk assessment of mycotoxins present in breakfast cereals consumed by children from Lisbon region, *Portugal Food Chem Toxicol.*, *86*, 274– 281.

[78] Polychronaki, N., Turner, P. C., Mykkänen, H., Gong, Y., Amra, H., Abdel-Wahhab, M. & El-Nezami, H. (2006). Determinants of aflatoxin M1 in breast milk in a selected group of Egyptian mothers *Food Addit Contam. 23*, 700– 708.

[79] Muñoz, K., Wollin, K., Kalhoff, H. & Degen, G. (2013). Occurrence of the mycotoxin ochratoxin a in breast milk samples from Germany *Gesundheitswesen*, *75*, 194–197.

[80] Yazdanpanah, H., Mohammadi, T., Abouhossain, G. & Cheraghali, A. M. (2005). Effect of roasting on degradation of aflatoxins in contaminated pistachio nuts. *Food Chem Toxicol*, *43*, 1135–1139.

[81] Saalia, F. K. & Phillips, R. D. (2011). Reduction of aflatoxins in peanut meal by extrusion cooking in the presence of nucleophiles. *LWT Food Sci Technol*, *44*, 1511–1516.

[82] Chen, J. M., Zhan, H. H., Yang, M. H. & Jin, Y. (2005). Recent advances in the determination methods of aflatoxins. *Zhongguo Zhong Yao Za Zhi*, *30*(24), 1890-4.

[83] Zheng, M. Z., Richard, J. L. & Binder, J. (2006). A review of rapid methods for the analysis of mycotoxins. *Mycopathologia*, *161*, 261–273.

[84] Maragos, C. M. (2004). Emerging Technologies for Mycotoxin Detection. *Journal of Toxicology Toxin Reviews*, *23*, 317–344.

[85] Czeh, A., Mandy, F., Feher-Toth, S., Torok, L., Mike, Z., Koszegi, B., et al. (2012). A flow cytometry based competitive fluorescent microsphere immunoassay (CFIA) system for detecting up to six mycotoxins. *Journal of Immunological Methods*, *384*, 71–80.

[86] Krska, R. & Molinelli, A. (2009). Rapid test strips for analysis of mycotoxins in food and feed. *Anal Bioanal Chem*, *393*, 67–71.

[87] Goryacheva, Y., De Saeger, S., Eremin, S. A. & Van Peteghem, C. (2007). Immunochemical methods for rapid mycotoxin detection: Evolution from single to multiple analyte screening: A review. *Food Additives and Contaminants*, *24*(10), 1169–1183.

[88] Quian, G. S., Ross, R. K., Yu, M. C., Yuan, J. M., Gao, Y. T., Henderson, Be. et al. (1994). A Foolow-Up Study of Urinary Markers of Aflatoxin Exposure and Liver Cancer Risk in Shanghai, People's Republic of China. *Cancer Epidemiology, Biomarkers & Prevention*, *3*, 3-10.

[89] Alberts, J. F., Gelderblom, W. C. A., Botha, A. & van Zyl, W. H. (2009). Degradation of aflatoxin B1 by fungal laccase enzymes. *Int J Food Microbiol*, *135*, 47–52.

[90] Hedayati, M. T., Pasqualatto, A. C., Warn, P. A., et al. (2007). Aspergillus flavus: human pathogen, allergen and mycotoxin producer. *Microbiology*, *153*, 1677-1692.

[91] Warth, B., Braun, D., Turner, P. & Marko, D. Biomonitorning of Mycotoxins in Human Breast milk: Current State and Future Perspectives, Vienna, 2016.

[92] Izvještaj o stanju u oblasti bezbijednosti hrane u Bosni i Hercegovini u (2015). godini, Mostar, 2016. [Report on the state of food safety in Bosnia and Herzegovina in 2015, Mostar, 2016.].

[93] Stroka, J. & Anklam, E. (2002). New strategies for the screening and determination of aflatoxins and the detection of aflatoxin-producing moulds in food and feed. *Trends in analytical chemistry*, *21*(2), 90-95.

[94] Reddy, K. R. N., Farhana, N. I. & Salleh, B. (2011). Occurrence of Aspergillus spp. and Aflatoxin B1 in Malaysian Foods Used for Human Consumption. *Journal of Food Science*, *76*(4), 99-104.

[95] Leong, Y. H., Ismail, N., Latif, A. A. & Ahmad, R. (2010). Aflatoxin occurrence in nuts and commercial nutty products in Malaysia. *Food Control*, *21*, 334–338.

[96] Zheng, Z., Hanneken, J., Houchins, D., King, R. S., Lee, P. & Richard, J. L. (2005). Validation of an ELISA test kit for the detection of ochratoxin A in several food commodities by comparison with HPLC. *Mycopathologia*, *159*, 265–72.

[97] Ratnavathi, C. V., Komala, V. V., Kumar, B. S. V., Das, I. K. & Vishnu Patil, J. V. (2012). Natural occurrence of aflatoxin B1in sorghum grown in different geographical regions of India. *J Sci Food Agric*, *92*, 2416–2420.

[98] Reddy, K. R. N., Farhana, N. I. & Salleh, B. (2011). Occurrence of Aspergillus spp. and Aflatoxin B1 in Malaysian Foods Used for Human Consumption. *Journal of Food Science*, *76* (4), 99-104.

[99] Leong, Y. H., Ismail, N., Latif, A. A. & Ahmad, R. (2010). Aflatoxin occurrence in nuts and commercial nutty products in Malaysia. *Food Control*, *21*, 334–338.

[100] Zheng, Z., Hanneken, J., Houchins, D., King, R. S., Lee, P. & Richard, J. L. (2005). Validation of an ELISA test kit for the detection of ochratoxin A in several food commodities by comparison with HPLC. *Mycopathologia*, *159*, 265–72.

BIOGRAPHICAL SKETCH

Mufida Aljicevic

Affiliation: Department of Medicine, University of Sarajevo, Sarajevo, Bosnia and Herzegovina

Education:

1986 MD Department of Medicine, University of Sarajevo

2000 Spec. microbiology. Department of Medicine, University of Sarajevo

2004 Master. Medical microbiology-bacteriology, Department of Medicine, University of Sarajevo

2010 PhD Medical microbiology- mycology, Department of Medicine, University of Sarajevo

Business Address: Cekalusa 90, Sarajevo 71000, Bosnia and Herzegovina

Research and Professional Experience:

2008 School of Medicine, South Carolina, SAD – Program of the Bureau of Educational and Cultural Affairs of the United States Department of State /"teaching staff" JFDP/

2013 Institute of Public Health, Universita Cattolica Del Sacro Cuore-School of Medicine "A. Gemelli" Rome, Italy /"Post Doctorate" ERAWEB/, research projects:

- Health Technology Assessment of the new meningococcal vaccine against serogroup B
- Systematic revew and meta-analysis of the effectiveness of heptavalent conjugated pneumococcal vaccine.

2015 Department of Medical Science University of Turin, Italy "teaching staff" ERAWEB/, research projects:

- Antimicrobial-coupled nanodroplets for skin and soft tissue infections: in vitro models.
- Advanced tecnologies in orthopedic surgery: innovative prosthetic biomaterials with enhanced osteointegration and antimicrobial properties.
- Influence of essential oil upon PMN intracellular killing.
- Screening programme for the prevention of cervical cancer. reasearch and diagnostic approaches in cancer relted infections (i.e., Papillomavirus).

Professional Appointments: Professor, Department of Medicine, University of Sarajevo

Publications from the Last 3 Years:

[1] Aljičević M, Redžić I, Avdagić Mevlida, Gondžetović N, Redžić A (2014). In vitro isolation of stem cells by bone-tissue engineering. *Folia Medica Facultatis Medicinae Univeristas Saraviensis*, 49(1); 48.

[2] Aljicevic M, Karcic E, Bektas S, Karcic B (2015). Representation of Streptococcus Pneumoniae in Outpatient Population of Sarajevo Canton. *Med Arch*, 69(3); 177-180.

[3] Karcic E, Aljicevic A, Bektas S, Karcic B (2015). Antimicrobial Susceptibility/Resistance of Streptococcus Pneumoniae. *Mater Sociomed*, 27(3); 180-4.

[4] Chiara de Vaure, Maria Lucia Specchia, Silvio Capizzi, Mufida Aljicevic, Milos Dujovic, Admir Malaj and Walter Ricciardi 2015. Effectiveness of 7-valent pneumococcal conjugate vaccine: A meta-analysis of post-marketing studies. *World J Meta-Anal*, 3(3);151-162.

[5] Bektaš S, Obradović A, Aljičević M, Numanović F, Hodžić D, Sporišević L (2016). The frequency of community-acquired methicillin-resistant Staphylococcus aureus (CA-MRSA) among samples in institute for public health in Canton Sarajevo. *Mater Sociomed*, 28(1); 61-65.
[6] Rebić V, Budimir A, Aljičević M, Mahmutović Vranić S, Rebić D (2016). Typing of Methicillin Resistant Staphylococcus Aureus Using DNA Finferprints by Pulsed-field Gel Electrophoresis. *Acta inform med*, 24(4); 248-252.
[7] Abduzaimović A, Aljičević M, Rebić V, Mahmutović Vranić S, Abduzaimović K, Šestić S (2016). Antibiotic Resistance in Urinary Isolates of Escherichia coli. *Mater Sociomed*, 28(6); 416-419.
[8] Mahmutović Vranić S, Zatić N, Rebić V, Aljičević M, Abduzaimović A (2017). The Most Frequent Isolates from Outpatients with Urinary Tract infection. *Mater Sociomed*, 29(1); 17-20.

Damir Ljuboja

Affiliation: Department of Medicine, Harvard Medical School, Boston, Massachusetts, US

Education:

BS Biology, BA Humanities, University of Texas at Austin, 2014
MD Candidate, Harvard Medical School (estimated 2020)
MBA Candidate, Harvard Business School (estimated 2019)

Research and Professional Experience:

Harvard Business School (Cambridge, MA) – *Lead Researcher*
January 2015 – May 2016

- Researched career trajectories of all physicians who obtained MBA degrees at Harvard Business School
- Worked with colleagues at HMS, HBS, Stanford, and Google Ventures to analyze trends in data

Global Health Delivery Project at Harvard University (Cambridge, MA) – *Fellow* May 2015 – September 2015

- Created and managed Expert Panel using GHDonline platform on the topic of digital clinical resources, their use abroad, and limitations to uptake for an estimated readership of 8,000-10,000 healthcare professionals
- Designed and executed survey study in Sarajevo, Bosnia and Herzegovina

Harvard Medical School/Children's Hospital Boston (Boston, MA) – *Research Fellow* June 2011 – July 2012

- Worked in Stuart Orkin's hematology/oncology laboratory on the identification of novel enhancer regions on the Dlk1-Dio3 imprinted gene cluster for improvement of induced pluripotent stem cell generation techniques
- Performed cloning procedures, chromatin conformation capture studies, luciferase assays, FACS flow cytometry, generated microRNA profiles of polycomb knockout embryonic stem cells, and conducted H3K4me3 ChIP
- Funded by the Howard Hughes Medical Institute

Honors: New England Journal of Medicine Scholar (1 of 40 undergraduates internationally), 2014

Publications from the Last 3 Years:

Ljuboja D, Robbins B, Powers BW, Huckman R, Yeshwant K, and Jain SH (2016). When Doctors Go to Business School: Career Choices of Physician-MBAs. *American Journal of Managed Care*, 22(6); 196-198.

Das PP, Hendrix DA, Apostolou E, Buchner AH, Canver MC, Beyaz S, Ljuboja D, Kuintzle R, Kim W, Karnik R, Shao Z, Xie H, Xu J, De Los Angeles A, Zhang Y, Choe J, Leong Jia Jun D, Shen X, Gregory RI, Daley G, Meissner A, Kellis M, Hochedlinger K, Kim J, and Orkin SH (2015). PRC2 is required to maintain expression of the maternal Gtl2-Rian-Mirg locus by preventing de novo DNA methylation in mESCs. *Cell Reports*, 12; 1-15.

In: Focus on Aflatoxins Research
Editors: P. Jan and S. Sylvester
ISBN: 978-1-53612-569-6

Chapter 3

NATURAL APPROACHES FOR CONTROLLING AFLATOXINS IN AGRICULTURE

***Hsin-Bai Yin*[1], *Anup Kollanoor-Johny*[2] and *Kumar Venkitanarayanan*[1,*]**

[1]Department of Animal Science, University of Connecticut, Storrs, Connecticut, US

[2]Department of Animal Science, University of Minnesota, St. Paul, Minnesota, US

ABSTRACT

Aflatoxins (AF) are fungal toxic metabolites produced by *Aspergillus flavus* and *Aspergillus parasiticus* that frequently contaminate a variety of crops, including peanut, corns, and cottonseed. Aflatoxin contamination causes an estimated average economic loss of $1 billion annually to the US agriculture. Additionally, contamination of feed ingredients with AF is a major concern to animal agriculture due to their deleterious effects on livestock such as reduced animal performance and increased mortality. Moreover, the consumption of AF-contaminated animal products negatively affects public health due to their carcinogenic and hepatotoxic

* Corresponding author: kumar.venkitanarayanan@uconn.edu.

effects. The common practices for detoxification of AF in crops include prolonged heating under pressure and the use of chemicals such as oxidizing agents and ammonia. However, nutrient losses associated with heat processing and public health safety concerns on chemical residues have limited the applicability of these methods. Moreover, the inclusion of AF-binding adsorbents in feed to protect livestock from the harmful effects of AF has been shown to impair nutrient utilization and mineral absorption in livestock. Thus, it is critical to develop effective strategies for controlling AF in crops, feed ingredients and livestock. Extensive studies have identified several natural plant- and animal-derived compounds exhibiting potent antifungal or antitoxigenic properties against toxin producing molds. This chapter highlights the potential of various natural approaches, including atoxigenic or probiotic microorganisms, phytochemicals, and organic compounds in reducing AF as well as the mechanisms behind their antitoxigenic effect. In addition, their potential application in controlling AF contamination at pre- and post-harvest stages of agriculture is discussed.

1. INTRODUCTION

Aflatoxins (AF) are toxic metabolites produced by *Aspergillus flavus* and *Aspergillus parasiticus. A. flavus* and *A. parasiticus* are closely related fungi that are known to contaminate numerous crops in the field, during harvest, in storage, and during processing, and can grow as unspecialized saprophytes on crop debris in soil (Wicklow and Donahue, 1984). The first outbreak of aflatoxin poisoning was reported in turkeys, which was identified as Turkey "X" disease in England in 1960s, causing the death of 100,000 turkeys (Blount, 1961; Van der Zijden, 1962). The Turkey "X" disease led to the discovery of AF from groundnut meal contaminated by *A. flavus* (Hesseltine, 1979), and the term "aflatoxin" came from "*Aspergillus flavus* toxin" (Yu et al., 2004). Since the discovery of the AF, researchers have been intensively studying their toxicity, carcinogenicity, mutagenicity, and teratogenicity in humans and animals (Ellis et al., 1991).

Among the 16 different types identified, Aflatoxin B1, B2, G1, and G2 are the major toxins that deleteriously affect agriculture (Goldblatt, 1971). Aflatoxin B1 is one of the most potent hepatocarcinogens, and its active metabolite, AFB1 8, 9, -epoxide, metabolized by cytochrome P450 in liver, binds to proteins and forms adducts such as aflatoxin B1-lysine in albumin. Furthermore, this active metabolite binds to guanine residues in DNA, forming guanyl-N^7 adducts, and initiates hepatocarcinoma (Hsu et al., 1991). Due to its carcinogenic properties, AFB1 has been listed as a group I human carcinogen by the International Agency for Research on Cancer (Yunus et al., 2011).

Link (1809) first used the name *A. flavus* to describe a mold species as well as a group of closely related species. In general, *A. flavus* is known as a velvety, yellow to green or brown mold with a golden to red-brown color (Hedayati et al., 2007). *A. flavus* isolates can be divided into two phenotypic types based on the characteristics of the sclerotia that are produced; the S strain produces numerous small sclerotia (average diameter, 400 mm) and the L strain produces fewer but larger sclerotia (Cotty, 1989). Both *A. flavus* S and L strains produce AFB1 and B2, but *A. flavus* S strains and *A. parasiticus* may also produce AF G1 and G2 (Cotty, 1989; Horn et al., 2003; Yu, 2005).

In general, warm and humid conditions (30°C and water activity of 0.99) at pre-harvest and post-harvest levels, and during transport or storage time, promote toxigenic *Aspergillus* spp. colonization, followed by AF production and accumulation in crops and feedstuffs (Cotty and Jaime-Garcia, 2007; Rawal et al., 2010). The economic loss due to mycotoxin contamination in agriculture and the cost of research and monitoring activities to control mycotoxins are estimated to range between $500 million and $1.5 billion annually (Abarca et al., 1994; Robens and Cardwell, 2003).

The common practices for detoxification of AF in crops include prolonged heating under pressure and the use of chemicals such as oxidizing agents and ammonia. However, nutrient losses associated with heat processing and public health safety concerns of chemical residues have limited the applicability of these methods. Moreover, the inclusion of aflatoxin-binding adsorbents in feed to protect livestock from the harmful effects of aflatoxins has been shown to impair nutrient utilization and mineral absorption in livestock (Kubena et al., 1993; Edrington et al., 1997). Thus, extensive research has been conducted to develop effective strategies for controlling AF in crops, feed ingredients, and livestock using natural approaches.

In this chapter, we discuss the potential of various natural approaches, including atoxigenic or probiotic microorganisms, phytochemicals, and organic compounds for controlling AF in agriculture and the mechanisms behind their antitoxigenic effect.

2. Aflatoxin Contamination in Agriculture

2.1. In Crops

A. flavus and *A. parasiticus* are commonly found in a wide range of climate zones, especially between latitudes 16° and 35° in warm areas (Klich, 2007). Their fungal sclerotia are able to survive in soil under severe environmental conditions and produce conidia dispersed by air or soil movement, leading to a population increase under hot and drought weather conditions (Wicklow et al., 1993; Payne, 1998). Conidia that are distributed through air are capable of infecting crops such as maize, tree nuts, peanut seed and cottonseed (Horn and Pitt, 1997; Cotty, 2001). The colonization of these fungi on crops could be enhanced with favorable environmental conditions (e.g., hot and dry climate) and damage caused by insects and birds (Horn and Pitt, 1997; Payne, 1998). Insect infestations of crops such as the lesser cornstalk borer (*Elasmopalpus lingosellus* Zeller), nitidulid beetles (*Nitidulid*), and rice weevil (*Sitophilus zeamais*) are associated with enhanced *Aspergillus* infections (LaPrade and Manwiller, 1977; Lussenhop and Wicklow, 1990).

Susceptibility of crops to *A. flavus* and *A. parasiticus* infections varies. For example, *A. flavus* infects a wide range of plant hosts, whereas *A. parasiticus* is generally limited to ground crop hosts due to preferences for suitable growth temperature (Horn et al., 2009). *A. flavus* as an opportunistic pathogen of agricultural crops infects oil-containing crops such as maize, peanut, and cottonseed, and it is considered as the dominant species compared to *A. parasiticus* on maize and cottonseed. *A. parasiticus* on the other hand, appears to be adapted to a soil environment, hence prominent in peanuts (Horn et al., 1998).

2.2. In Animal Feed

Cereals are major AF vectors because they are consumed by both humans and animals (Dhanasekaran et al., 2011). Approximately 25-40% of cereals are contaminated with mycotoxins in the world, AF as the most common contaminant among the mycotoxins (Pittet, 1998). *A. flavus* and *A. parasiticus* infect and produce AF in cereals such as corn, peanuts, and cottonseeds that are commonly used as feed ingredients (Dhanasekaran et al., 2011). Aflatoxins unintentionally appear in animal feed as a result of the inadvertent use of AF-contaminated feed ingredients such as cereal grains. Once food-producing animals ingest AF-contaminated feed, toxin residues can be transferred to animal/poultry meat and eggs (CAST, 2003; Pandey and Chauhan, 2007; Aly and Anwer, 2009). In addition, AF M1, an intermediate metabolite of AF, was found at the maximum level in milk two days after the exposure to AF-contaminated feed with a transfer rate of AF between 0.3-2.2% from animal feed to milk.

2.3. Economic Impact

Aspergillus colonization does not necessarily reduce the yield of affected crops, but causes economic losses due to AF contamination. The Food and Agriculture Organization (FAO) reported that 25% of the world's grains are contaminated by mycotoxins, and AF contamination is the most common among them. Worldwide, AF are considered a major public health problem, especially in developing countries, where long-term food storage is often inadequate for high heat and humidity, which encourage mold growth in grains and other ingredients (Eaton and Gallagher, 1994). The economic impacts attributed to AF are incurred directly by losses in crops, livestock, and dairy, and indirectly by a recurring expenditure in quality-control programs, research and education, lower foreign exchange earnings, and increased storage and packaging costs of vulnerable commodities (Mishra and Das, 2003).

Mycotoxin contamination in agriculture has caused an estimated average economic loss at approximately $1 billion annually in the United States (Vardon et al., 2003), and among all the mycotoxins, AF contamination is the most serious worldwide (CAST, 2003). According to Rubens and Cardwell's report, AF contamination resulted in an annual economic loss of approximately $25 million on peanut production in the state of Georgia, $2-15 million losses on maize production in Texas, Mississippi, and Arizona, $3.8 million on walnut production in California, and $2.3-4.7 million in California (Rubens and Cardwell, 2005). Although the impact of economic loss is not available from developing countries in Asia and Africa, it is likely that they are more severely affected (Yu et al., 2005).

3. Health Impacts of AF

3.1. In Humans

Once AF enter the host, it is first absorbed in the small intestine, especially in the duodenum (Gratz et al., 2005). However, because enzymes present in the small intestine have a low affinity for AF (Guengerich et al., 1996), the majority of the absorbed AF is metabolized in the liver, where they initiate bioactivation (Eaton and Gallagher, 1994). During the biotransformation of AF in liver, AFB1 becomes an intermediate form, AFB1 8,9-epoxide, exerting its hepatocarcinogenic effects by binding with DNA, RNA, and proteins in the liver (Eaton and Gallagher, 1994; Leeson et al., 1995; Bedard et al., 2006).

Numerous epidemiological studies in humans naturally exposed to AF-contaminated foods could be found in the literature. Aflatoxicosis in humans can be classified into acute and chronic conditions based on the concentration of consumed AF and the period of time that the hosts are exposed to the toxin (Williams et al., 2004). Acute aflatoxicosis, resulting from consuming a high-concentration of AF over a short period of time often results in outbreaks in humans. For example, Kenya experienced the most serious recent outbreak of AF contamination in 2004 with 317 cases and 125 reported deaths (Azziz-Baumgartner et al., 2005). Low-dose AF consumption over an extended exposure time results in chronic aflatoxicosis, which can deleteriously affect human immune system, affect growth, and potentiates liver cancer. Approximately 4.5 billion people are estimated at risk of chronic exposure to AF-contaminated food (Hamid et al., 2013).

Several epidemiological studies have supported the correlation between the dietary AF intake and the incidence of human hepatocellular carcinoma, where AFB1 is the most hepatocarcinogenic mycotoxin and the main contributor to the high rate of carcinoma (Hamid et al., 2013). One of the mechanisms of AF-induced liver cancer is from a mutation in the tumor suppressor gene, *p53*, in the liver (Hsu et al., 1991). The rate of hepatocellular carcinoma was found to be increased by at least 30-fold in the presence of both AF and the hepatitis B or C virus infection, creating severe health problems in people living in developing countries, where both AF and hepatitis viruses are common (Kuang et al., 2005; Groopman et al., 2008). Similarly, Ross et al. (1992) analyzed over 18,000 urine samples for the presence of the N^7-guanine adduct of AF. In their study, 22 urine samples from the subjects who developed liver cancer during the analysis were examined for the presence of hepatitis B virus surface antigen and AF exposure biomarker. Results revealed that AF exposure alone (hepatitis B antigen-negative) yielded a relative risk of about two; hepatitis B virus antigen positive status alone (AF exposure negative) yielded a relative risk of about five. However, combined exposure (AF and hepatitis B virus positive antigens) yielded a relative risk of over 60 (Ross et al., 1992), indicating deleterious health outcomes when these two conditions co-exist.

In order to assess chronic AF exposure, studies have focused on the development of accurate and applicable biomarkers of exposure to toxin such as AF-related urinary metabolites due to dietary AF intake

(Groopman et al., 1988; Groopman et al., 1992; Qian et al., 1994). Zhu et al. (1987) conducted a study in China and found a high correlation score of 0.65 between total dietary AFB1 intake and urinary AFM1 excretion in 32 households in the Guangxi region. Similarly, Wild and co-workers (1992) found a high correlation between dietary intake and urinary excretion of AF metabolites; AFB1-N^7-guanine adducts in urine represented the most reliable urinary biomarker of AF exposure (Groopman 1988; Qian et al., 1994). Albumin-lysine-AFB1 adduct is another biomarker of AF exposure; Hall and Wild (1994) found a 10-fold increase in urinary AF metabolites over a 4-day period and a 2-fold increase in albumin-AFB1 adducts in the same period. In addition, in the same case-control study that investigated the relationship between AF and liver cancer in China, albumin-AFB1 adduct concentration in peripheral blood was correlated with individual dietary AF intake (Hall and Wild, 1994).

3.2. In Poultry

Animal susceptibility to the acute effects of aflatoxicosis varies widely. The presence of relatively high levels of AF in feeds can lead to animal death; turkeys, rabbits, ducks, and swine are particularly susceptible (LD_{50} = 0.5, 0.30, 0.35, and 0.62 ppm, respectively, Pier, 1992). The variation in the sensitivity of various animal species towards AF is believed to be linked to differences in the toxin's metabolism and the types of metabolites formed (Emafo, 1976). In general, domestic turkeys (*Meleagris gallopavo*) and ducks (*Anas platyrhynchos*) are more sensitive to both acute and chronic toxicity of AFB1 than chickens (*Gallus gallus*), except during embryonic development (Giambrone et al., 1985; Klein et al., 2000; Rawal et al., 2010). During chicken embryonic development, a LD_{50} dose of AFB1 to chicken embryo could be as low as 0.3 mg/kg body weight (Cullen and Newberne, 1993; Leeson et al., 1995). Even when toxin exposure does not cause mortality or morbidity, aflatoxicosis contributes directly and indirectly to losses in livestock and poultry industry.

It is well known that the liver is the primary organ of AF bio-activation and detoxification (Giambrone et al., 1985; Klein et al., 2000; Rawal et al., 2010). Micco and colleagues (1988) reported that when chickens were exposed to an AF contaminated diet, AFB1, AFM1, and aflatoxicol were detected in liver, kidneys, and thigh muscles. Longer exposure to AF contaminated feed leads to an increase in the relative weight of liver and causes pale or yellowed pigmentation in poultry (Verma et al., 2004; Yarru et al., 2009; Lee et al., 2012). At cellular level, increased vacuolation of AFB1-exposed hepatocytes allows high levels of lipids to accumulate (Sims et al., 1970; Giambrone et al., 1985; Oliveira et al., 2002). Steatosis is, therefore, responsible for the changes in liver color and size during aflatoxicosis.

Modern poultry is developed to gain more weight by utilizing less feed in a shorter time, which make them become more susceptible to AF contamination (Havenstein et al., 1994; Dozier et al., 2008). For example, AF might result in more profound negative effects in broiler chickens with more efficient nutrient conversion demanding faster hepatic metabolism. Both acute and chronic aflatoxicosis in animals cause AFB1-induced liver damage, including focal necrotic hepatocytes or hemorrhages (Newberne and Butler, 1969; Giambrone et al., 1985; Klein et al., 2002). Acute damage initiates inflammatory responses and results in leukocyte infiltration and proliferation in the liver (Cova et al., 1990). Additionally, short-term exposure to high doses of AF causes morbidity and mortality due to extensive liver damage (Rawal et al., 2010). In poultry, chronic AFB1 consumption is mutagenic and causes remodeling of liver tissues. Hyperplasia of bile duct epithelial cells or development of oval cells, followed by periportal fibrosis and nodular tissue regeneration has been reported (Cova et al., 1990; Ortatatli and Oguz, 2001).

Although AF are most often noted for the ability to induce liver cancer at very low doses, these toxins can cause several problems of economic importance during poultry production. AFB1 and its metabolites can be transferred from the laying hen into the albumin and yolk of the egg (Oliveira et al., 2000; Oliveira et al., 2003). Signs of aflatoxicosis in chicks and poults have been noted if they are exposed to AFB1 during development (Khan et al., 2014). Transfer of AF into embryonated eggs is a concern for poultry producers, because many studies with *in ovo* AFB1 injections confirmed the risk of AFB1 exposure to embryos. For example, *in ovo* exposure of chickens and turkeys to AFB1 caused DNA damage in the embryonic liver and increased embryo mortality (Dietert et al., 1985; Edrington et al., 1995; Celik et al., 2000; Oznurlu et al., 2012). When laying hens were fed with AFB1 contaminated feed to simulate the natural route of embryonic exposure, AF caused reduced hatchability and compromised cellular and humoral immune functions after hatch (Qureshi et al., 1998; Khan et al., 2014). In addition, embryonic AFB1 exposure can lead to morphological defects (Edrington et al., 1995) such as abnormal area opaca cells (Celik et al., 2000; Sur et al., 2003), skeletal defects in the tibia growth plate (Oznurlu et al., 2012), and inhibition of bursal follicle development (Celik et al., 2000; Sur et al., 2003), which can consequently reduce embryo viability adversely affecting the progeny. Moreover, embryonic exposure of AFB1 causes immunosuppression, thereby increasing the incidence of infectious disease in young poultry and detrimentally affecting their health and productivity.

Dietary exposure to AF leads to decrease in performance, including decreased body weight gain and absolute body weight in both chickens and turkeys (Giambrone et al., 1985; Quezada et al., 2000; Pandey and Chauhan, 2007). In addition, aflatoxicosis reduces feed intake and decreases efficiency of nutrient usage, resulting in increased feed conversion ratio (Verma et al., 2004; Yarru et al., 2009; Lee et al., 2012). For example, Dersjant-Li and coworkers (2003) reported that each mg of AFB1/kg diet decreased growth performance of broilers by 5%. Similarly, Miazzo et al. (2000) reported a reduction of 11% in body weight gain when 2.5 mg AFB1/kg diet was fed to broilers from 21 to 42 days of age compared to non-contaminated controls. Similarly, when layer hens were fed AFB1 in their diet, age to maturity was increased and egg production decreased (Azzam and Gabal et al., 1998; Garlich et al., 1973; Howarth et al., 1976; Khan et al. 2014).

3.3. In Livestock

Soon after the outbreak of Turkey "X" disease, researchers discovered that AF contamination was responsible for another outbreak of hepatocellular carcinoma in hatchery-reared rainbow trout, a species for which tumor rates were very low (Halver, 1969). Since the discovery of the AF in 1960s, studies conducted during the last five decades have investigated the negative effects of AF on various livestock, including effects on animal performance and metabolism, metabolism of the toxin, and carryover of toxic residues in animal products (Diaz, 2005).

Aflatoxicosis in livestock resulting from the ingestion of AF contaminated feed negatively affects production parameters, causing severe economic loss to the industry. However, ruminants are more resistant to most of the mycotoxins than monogastrics such as pigs and poultry, due to the detoxification of toxins by gastrointestinal microorganisms. AF are poorly degraded in the rumen, with less than 10% degraded for concentrations from 1-10 ppm (Westlake et al., 1989). Symptoms of acute aflatoxicosis in livestock include lethargy, ataxia, rough hair coat, and enlarged fatty livers, whereas chronic AF exposure results in reduced feed efficiency, milk production, and decreased appetite (Whitlow et al., 2010). Guthrie (1979) reported decreased reproductive efficiency of cows after feeding 120 ppb AF-contaminated feed. AF in feed significantly reduced the production of milk in cows (Applebaum et al., 1982).

4. Traditional Treatments for Controlling AF in Agriculture

Aflatoxin contamination may occur in the field before harvest, during harvesting, or during storage and processing, thus methods for the prevention of contamination can be divided into pre-harvest and post-harvest strategies. Because of the recognition of AF as a potent human carcinogen, federal regulation of allowable amounts of the toxin in food and feed is nearly universal. The Food and Drug Administration (FDA) in the United States regulates the amount of allowable AF contamination as 20 ppb in crops or 0.5 ppb in milk for humans. In Europe, foods containing more than 2 ppb of AFB1 or 4 ppb of AF are rejected, whereas the limit of AF contamination is 30 ppb in India (Van Egmond and Jonker, 2004).

The recommendations for the reduction of AF in agriculture include Good Agricultural Practice (GAP), Good Manufacturing Practice (GMP), and the use of Hazard Analysis Critical Control Point (HACCP) (Mejía-Teniente et al., 2011). Hazard Analysis Critical Control Point (HACCP) system involves a science-based analysis of potential hazards involved in the production of foods and is designed to critically evaluate the effectiveness of controls at each major point involved in the processing steps. The implementation of HACCP principles to minimize AF contamination has been successfully applied in Southern Africa in commercially produced peanut butter (Mejía-Teniente et al., 2011). Current control methods that reduce the incidence of AF contamination in agriculture also include development of resistant strains, physical and chemical methods, and toxin binders.

4.1. Development of Resistant Strains

One strategy to control AF contamination at the pre-harvest level is based on the identification of crop lines resistant to insects or harsh environmental conditions such as drought, because insects and drought stress damage crops, thereby favoring AF contamination (Lynch et al., 2003; Cotty and Jaime-Garcia, 2007). Several researchers have focused on identifying fungus-related and AF-resistant genes and proteins that are important for defense against *A. flavus* invasion and/or AF bio-synthesis (Chen et al., 2006; Guo et al., 2006; Chen et al., 2010). Therefore, genetic modifications of crops such as corn, cottonseed, and peanut may gain resistance to AF contamination by being transferred with the genes that are responsible for AF inhibition enzymes (Cary et al., 2000, Mishra and Das, 2003).

4.2. Physical Methods

The continued cultivation of crops such as peanuts on the same land may lead to a built-up of high populations of *A. flavus* and *A. parasiticus* in the soil, which will increase the probability of infection and subsequent AF contamination. Studies have been carried out to demonstrate the effect of crop rotation on AF contamination; however, in semi-arid environments, populations of *Aspergillus* may be very high, and crop rotations may have little influence on the fungal activity.

There are several physical methods for detoxification of AF at post harvest level such as prolonged heating with pressure; however, certain nutrients also could be destroyed during the process. Other physical methods also include manually sorting out of contaminated grains (detected by fluorescence) and removal of contaminated grains or kernels via flotation. These processes have at least two drawbacks: (1) high cost of removing and disposing of the contaminated materials, and (2) difficulty achieving complete removal contaminated materials without wasting significant portions of uncontaminated products (Yiannikouris and Jouany, 2002).

4.3. Chemical Methods

A variety of chemicals have also been screened for their ability to react with AF such as methanol, oxidizing agents (hydrogen peroxide, ozone), reducing agents (bisulphites), acids, and bases (ammonia, caustic soda) (Samarajeewa et al., 1990). Some mycotoxins can be destroyed chemically with calcium hydroxide, monoethylamine, ozone or ammonia, where ammoniation degrades 95-98% of the AF (Moss et al., 1998). For example, de Alencar et al. (2012) demonstrated the fungicidal and detoxifying effects of ozone on AF in peanuts. In their research, results showed that ozone was effective in inactivating populations of aflatoxigenic species in peanuts and a reduction in the concentration of total AF and AFB1 of approximately 30% and 25%, respectively, was observed for kernels exposed to ozone at 21 mg/L for 96 h. However, loss of nutritional and sensory qualities of the product during the chemical process, as well as, the expensive equipment required for these techniques have encouraged an emphasis on biological methods of control (Teniola et al., 2005).

4.4. Inorganic Toxin Binders

Several interventions have been developed to control AF contamination in feed to reduce aflatoxicosis in animals and subsequent transfer of AF residues from food-producing animals to humans. To minimize the considerable economic losses caused by AF contaminated crops (Henry et al., 1999), various non-nutritive adsorbents have been employed for reducing or inactivating AF in feeds. For instance, supplementation of an inorganic toxin binder such as clay in the feed is one of the widely used approaches to control AF in the feed industry. The clay selectively binds tightly to AF to prevent their absorption in the gastrointestinal tracts and the clay-AF complex is eliminated from the body (Hell et al., 2008). Evidence suggests that AF may react at multiple sites on the clay particle, especially interlayer regions, edges, and basal surfaces (Mishra and Das, 2003).

Hydrated sodium calcium aluminosilicates (HSCAS) is another toxin binder, which reduces AF absorption by binding with the β-carbonyl portion of AF molecules, thereby effectively reducing aflatoxicosis (Scheideler, 1993; Ledoux et al., 1999). Hydrated sodium calcium aluminosilicates (HSCAS) have also been demonstrated to act as an AF enterosorbent that tightly and selectively binds the toxin in the gastrointestinal tract of animals, thus decreasing their bioavailability and consequently alleviating aflatoxicosis (Scheideler, 1993; Phillips, 1999). Such adsorbents act more as prophylactics than curative remedies. However, there may be certain risk factors for their inclusion in diet before proper testing, since several adsorbents have been shown to impair nutrient utilization (Kubena et al., 1993) and mineral absorption in animals (Edrington et al., 1997). In addition, these inorganic absorbents are not extensively biodegradable and are associated with the risk of contaminants??? incomplete?

5. Natural Approaches for Controlling AF in Agriculture

In principle, there are three possible ways to avoid harmful effects caused by AF: (1) prevention of AF producing fungi at pre-harvest stage; (2) detoxification of AF-contaminated food and agricultural commodities at post-harvest level; (3) inhibition of absorption of AF from consumed food in the digestive tract to reduce aflatoxicosis. Although AF contamination occurs at both pre- and post-harvest levels, post-harvest AF contamination of seeds is the major problem because of improper storage practices, primarily under excessive moisture, and insect activity (Cotty, 1997).

For a detoxification method to be acceptable, it must be efficient, safe and cost effective while safeguarding nutritional quality (Hell et al., 2008). As of now, numerous studies have identified several natural plant-derived compounds exhibiting potent antifungal or antitoxigenic properties against toxin producing molds. Here, we have summarized the efficacy of several natural approaches, including atoxigenic or probiotic microorganisms, phytochemicals, and organic compounds in reducing AF contamination in agriculture.

5.1. Biological Control Methods

Utilization of non-aflatoxigenic *A. flavus* strains (or called atoxigenic strains; strains that cannot produce AF) as a biocontrol agent to inhibit the growth of aflatoxigenic *A. flavus* in the field has been investigated. Mechanism of this bio-control is through competitive exclusion or inhibition of toxigenic *A. flavus* by non-toxigenic strains for nutrient substrates (Cleveland et al., 2003; Pitt and Hocking, 2006). Large-scale development of atoxigenic strains has been undertaken since 1998 by the Arizona cotton industry (Antilla and Cotty, 2001). To ensure the effectiveness of atoxigenic *A. flavus* strains to control AF contamination on crops, they must be applied at a time and in a manner, that allows successful competition with AF-producers (Cotty et al., 1994). This means atoxigenic strains should be applied when the AF concentrations on the crops are still comparatively low (Cotty et al., 1994; Cotty, 1997).

Two biological control agents using non-toxigenic *A. flavus* strains registered by the U.S. Environmental Protection Agency are currently commercially available. Afla-Guard is one of the agents that are composed of hulled barley coated with conidia of non-toxigenic *A. flavus* strain NRRL21882, which is registered for use on peanuts and corn (Dorner, 2010). The other biocontrol agent is *A. flavus* AF36, primarily used for reducing AF on cottonseed. Here, *A. flavus* strain NRRL21882 lacks the AF bio-synthesis cluster from hexanoate synthase alpha gene *hexA* to the telomeric region, and AF36 is defective in the AF polyketide gene *pksA*. Thus, both stains are not able to produce AF (Ehrlich and Cotty, 2004). The first commercial use of Afla-Guard in the United States, resulted in an AF reduction averaging 85% in farmers' stock peanuts and as high as 98% in shelled stock (Dorner, 2010). The same strategy was applied to prevent the AF contamination in peanuts from Australia and Argentina (Pitt and Hocking, 2006). Recently, studies have been conducted to determine the most effective way to incorporate biological control method into standard agricultural practice using multi-agency, multi-state, and large-scale field trials (Abbas et al., 2008).

Similarly, small-scale experiments have also been conducted on competitive exclusion using *Bacillus thuringiensis*, the bacterium that produces toxins. *B. thuringiensis* has not only been assessed for biological control properties against insects for decades, but also been studied for its detrimental effects on fungal colonization and mycotoxin contamination (Zhakharian et al., 1978). However, results were not consistent in terms of the efficacy on *B. thuringiensis* corn in reducing AF contamination. For example, Abbas et al. (2008) documented that AF contamination has been reduced in *B. thuringiensis* treated corn, whereas Buntin et al. showed no significant differences between non-*B. thuringiensis* and *B. thuringiensis* treated corn (Buntin et al., 2001). Scientific reports have indicated that tremendous reduction of AF levels in *B. thuringiensis* treated maize can occur when high levels of insect resistance are available (Cleveland et al., 2003).

Biocontrol based on competitive exclusion by non-toxigenic strain of *A. flavus* has also demonstrated that field application of the biocontrol agents exerted a carry-over effect and reduced AF contamination during storage (Dorner and Cole, 2002). At the post-harvest intervention, Kong et al. (2010) evaluated the efficacy of a strain of marine *Bacillus megaterium* isolated from the Yellow Sea of East China for controlling decay of peanut kernels caused by *A. flavus*, and the results showed that the antagonist had a significant effect on biocontrol effectiveness *in vivo* significantly reducing the biosynthesis of AF and expression of *aflR* gene and *aflS* gene (Kong et al., 2010).

5.2. Organic Toxin Binders

Another group of toxin binders belong to the category of organic absorbents such as activated charcoal, synthetic polymers, yeast cell walls, and bacterial cells. However, the effects of activated charcoal have been variable, and responses to charcoal with poultry also suggest that charcoal may not be as effective in binding AF as are clay-based binders (Rezaei et al., 2014). Research done by Diaz et al. (2005) showed that activated carbon did not significantly reduce AF residues in milk when the cows were exposed to AF contaminated diet, whereas another organic polymer of esterified glucan significantly reduced milk AF. Responses to charcoal with broilers (Edrington et al., 1997), turkey poults (Edrington et al., 1997), rats (Abdel-Wahhab et al., 1999), and mink (Bonna et al., 1991) also suggest that charcoal may not as effective in binding AF as are clay-based binders.

Yeast cell walls, particularly the cell wall of *Saccharomyces cerevisiae*, are an environmentally friendly alternative to inorganic adsorbents. In addition to the beneficial effects observed in counteracting AF, these organic binders would be efficient against a large range of mycotoxins, which make them more adapted to the most frequent cases of multi-contaminated feed (Shetty and Jespersen, 2006). Bacterial cell walls also have potential to bind AF, but limited research has been conducted.

5.3. Phytochemicals

Plant-derived essential oils are a group of natural, environmentally friendly antimicrobials that have traditionally been used as food preservatives and flavor enhancers. In the past decade, the use of plant-derived compounds has gained significant attention due to increasing concern over the safety of synthetic chemicals and emergence of antibiotic-resistant strains of microorganisms (Salamci et al., 2007). The antifungal and antitoxigenic properties of several plant oils have been identified (Burt et al, 2004). Phytochemicals are classified as generally recognized as safe (GRAS) for addition in food products by the FDA (Adams et al., 2004).

Significant research is being conducted to investigate the potential of natural phytochemicals in inhibiting the growth and AF production by *Aspergillus* species in food products and animal feed. Clove and cinnamon oils at 100 ppm were found to inhibit *A. flavus* growth and AF production *in vitro* (Sinha et al., 1993). In the same study, the researchers demonstrated that cinnamon oil at 1000 ppm exhibited maximum inhibitory action and reduced AF production by 78% on maize. Similarly,

Zataria multiflora was found to exert significant inhibitory effect on both *A. flavus* growth and AF production in cheese (Gandomi et al., 2009). Follow up mass spectrometric analysis revealed that the major component (71%) of *Z. multiflora* oil was carvacrol. The antitoxin effect of carvacrol on *A. parasiticus* was investigated by Razzaghi-Abyaneh and coworkers (2008). These researchers reported that carvacrol effectively suppressed the growth as well as the synthesis of AFB1 and G1 by *A. parasiticus in vitro*.

Additionally, an *in vitro* study conducted in the authors' laboratory confirmed the anti-toxigenic efficacy of carvacrol and *trans*-cinnamaldehyde against *A. flavus* and *A. parasiticus* in a broth system and in chicken feed during long-term storage (Yin et al., 2015a). Carvacrol is a major ingredient in oregano oil (*Origanum glandulosum*), whereas *trans*-cinnamaldehyde is a principal component in cinnamon bark (*Cinnamomum zeylandicum*). In the study, results indicated that carvacrol and *trans*-cinnamaldehyde reduced *A. flavus* and *A. parasiticus* growth and AF production in potato dextrose broth by at least 80%. In addition, all the concentrations of carvacrol and *trans*-cinnamaldehyde (0.4%, 0.8%, and 1%) decreased AF concentrations in the feed to levels below the FDA regulated limit (20 ppb) for up to 3 months. However, feed samples with no added carvacrol or TC yielded more than 30 ppb of AF.

Aflatoxins, as secondary metabolites of molds, have a very complicated biosynthesis pathway, which involves a cluster of at least 29 genes in a contiguous 75 kb region (Yu et al., 2004, Ehrlich et al., 2005). The mechanisms behind the anti-toxigenic effect of carvacrol and *trans*-cinnamaldehyde against *A. flavus* and *A. parasiticus* were further examined using real time quantitative PCR by Yin et al. (2015a). Results of the study revealed that carvacrol and *trans*-cinnamaldehyde significantly down-regulated the expression of critical genes associated with AF synthesis (*aflC*, *nor1*, *norA*, and *ver1* genes) in *Aspergillus* ($P < 0.05$), where *aflC* involved in the conversion of acetate to norsolorinic acid (NOR) plays a critical role in early AF biosynthesis pathway. Ehrlich and Cotty (2004) reported that a nonsense mutation of this gene resulted in the loss of AF synthesis in molds. Similarly, *nor1* and *norA* are involved in the conversion of NOR to averatin (AVN), where NOR is the key structural intermediate in the pathway (Bennett, 1981). Moreover, *nor1* has been used as a potential marker gene to discriminate between toxigenic or nontoxigenic strains (Mayer et al., 2003). *ver1* encodes an enzyme that converts versicolorin (VERA) to demethyl-sterigmatocyin (DMST), which is located in the last step of AF biosynthesis pathway before the intermediates start differentiating into different types of AF (AFB1, B2, G1, and G2).

During egg formation, AF residues could be transferred from laying hen to fertilized eggs, thereby resulting in decreased embryo viability and hatchability, and causing several organ malformations. Yin and co-workers (2015b) demonstrated the efficacy of carvacrol and *trans*-cinnamaldehyde in reducing AF-induced toxicity in chicken embryos using a fertilized egg model with *in ovo* injection of AF, as described by Celik et al. (2000) and Oznurlu et al. (2012). Results demonstrated that both carvacrol and *trans*-cinnamaldehyde significantly decreased AF-induced toxicity in chicken embryos. At the level of 75 ng of AF/egg, carvacrol and *trans*-cinnamaldehyde increased the survival of chicken embryo by ~55%. Moreover, carvacrol and *trans*-cinnamaldehyde significantly improved the growth of embryos (tibia length and weight) when compared to those injected with AF alone ($P < 0.05$).

Further, the efficacy of carvacrol and *trans*-cinnamaldehyde as feed supplements in controlling aflatoxicosis in chickens has also been reported (Yin et al., 2015c). A total of 240 chickens were fed with AF contaminated feed (~ 2.5 ppm) with or without supplementation of 0.75% carvacrol or *trans*-cinnamaldehyde for 5 weeks. Results showed that carvacrol and *trans*-cinnamaldehyde supplementation in AF-contaminated feed ameliorated AF-induced adverse effects in chickens. Histologic analysis revealed that carvacrol and *trans*-cinnamaldehyde reduced AF-induced toxic effects in the liver of birds, where phytochemical-treated chickens had decreased hepatocellular degeneration, necrosis and inflammation in the liver as compared to chickens fed with AF alone. To elucidate the protective mechanisms behind these phytochemicals, hepatic transcriptome analysis of chicken liver using RNA-seq identified several pathways and genes that are associated with aflatoxicosis in chickens, including hepatic steatosis, hepatic carcinoma, fatty acid metabolism, inflammation, and apoptosis being modulated with supplementation of these phytochemicals to AF-contaminated feed (Yin et al., 2015c).

5.4. Probiotic Bacteria

Lactic acid bacteria (LAB) and bifidobacteria, due to their GRAS status and use as probiotics, are of particular interest for reducing the bioavailability of AF. Karunaratne (1990) indicated that *Lactobacillus acidophilus, L. bulgaricus,* and *L. plantarum* could be used to inhibit the growth of AF-producing molds and degrade AF. *Bifidobacterium* is also reported to efficiently bind with AFB1 (Peltonen et al., 2000). The detoxification ability and toxin binding properties of AFB1 by six strains of probiotic bacteria were reported by Shah and Wu (1999). Results indicated that the detoxification ability of these organisms against AFB1 ranged between 20-50%, in which *Bifidobacterium infantis* strain 1912 demonstrated a significantly higher AF binding ability than other tested strains, while *B. pseudolongum* strain 20099 showed the least AF binding ability. Previous studies suggest that binding of AFB1 to lactic acid bacteria is a physical phenomenon associated with the cell wall of these microorganisms (Lillehoj et al., 1967; El-Nezami et al., 1998). The bacterial cell envelope appears to be the site of AF binding, with cell wall polysaccharides and peptidoglycan thought be the molecules of greatest importance (Peltonen et al., 2000; Haskard et al., 2001; Lahtinen et al., 2004).

In addition, the efficacy of *Berevibacillus laterosporus* isolated from quails for controlling aflatoxicosis in poultry was reported by Bagherzadeh Kasmani et al. (2012), where feeding AFB1 alone to the quail chicks resulted in a significant decrease in serum albumin, total protein, and glucose and cholesterol levels but a significant increase in serum uric acid, urea, and phosphorus ($P < 0.05$); however, treatment of birds on AFB1 with *B. laterosporus* restored these to their original levels ($P < 0.05$). Antibody titer against Newcastle disease virus was found to be lowest in the AFB1 group but highest in the *B. laterosporus* group ($P < 0.05$). It can be concluded that selected indigenous *B. laterosporus* is a promising probiotic with AFB1 removal potential. AFB1 binding by dairy strains of lactic acid bacteria and bifidobacteria was reported (Jiang et al., 2005; Teniola et al., 2005). Bovo and co-workers (2012) evaluated the ability of lactic acid bacteria to remove AFM1 in phosphate buffer saline (PBS). Heat-killed cells of *L. rhamnosus*, *L. delbrueckii* spp. *bulgaricus,* and *B. lactis* had the highest percentage (>33%) of AFM1 removal in PBS tests.

Table 1. Natural approaches for controlling aflatoxin in agriculture

Methods	Applications	Mechanisms	References
Biological control			
Non-aflatoxigenic *A. flavus* strains	Crops in the field Cotton barley peanuts	Inhibition of the growth of toxigenic *A. flavus*	Cleveland et al., 2003; Pitt and Hocking, 2006; Dorner, 2009
Bacillus thuringiensis; Bacillus megaterium	Crops in the field corns	Competitive exclusion	Zhakharian et al., 1979; Kong et al., 2010
Organic toxin binders			
Esterified glucan	Cow and milk	Toxin binding	Diaz et al., 2004
Yeast cell walls: *Saccharomyces cerevisiae*	Animal feed		Jans et al., 2012
Phytochemicals			
Clove, cinnamon	*In vitro*	Inhibition of the growth of toxigenic molds	Sinha et al., 1993
Zataria multiflora	Cheese		Gandomi et al., 2009
Carvacrol, *trans*-cinnamaldehyde	Chicken feed, chicken embryo, and broiler chickens	Regulation of aflatoxin synthesis genes in molds and pathways associated with aflatoxicosis in animals	Yin et al., 2015a, b, c
Probiotic bacteria			
L. acidophilus, L. bulgaricus, L. plantarum, B. infantis	*In vitro*	Toxin binding	Peltonen et al., 2001; Has- kard et al., 2001; Lahtinen et al., 2004
Berevibacillus laterosporus	Quail chicks		Kasmani et al., 2012

These studies demonstrate that use of phytochemicals, and atoxigenic and probiotic microorganisms offer a promising, viable and natural alternative to chemical antimicrobials in combating AF contaminations and promoting food safety.

Conclusion

Aflatoxins are toxic secondary metabolites produced by *A. flavus* and *A. parasiticus* that frequently contaminate crops and animal feedstuff at pre-harvest and post-harvest levels. Contamination of animal feed with AF is a major concern in animal industry due to their deleterious effects in animals. Additionally, the consumption of AF-contaminated animal by-products negatively affects public health due to their carcinogenic and hepatotoxic effects. Therefore, it is critical to develop effective strategies for controlling AF in agriculture. Research efforts in the past few" decades have provided important information to improve food safety from AF contamination. In this chapter, we have summarized several natural approaches that can be potentially used to control AF in food-producing animals, crops, and animal feedstuff. Antifungal and antitoxigenic mechanisms of action of these methods have been studied *in vitro*, much is still unclear about their efficacy in foods and animals. Although these natural interventions at the pre- and post-harvest levels showed promising results to control AF contamination, there is a need for continued research into agricultural industry practices to ensure food and feed safety.

References

Abarca, M. L., Bragulat, M. R., Castella, G., & Cabanes, F. J. (1994). Mycoflora and aflatoxin-producing strains in animal mixed feeds. *Journal of Food Protection*, *57*(3), 256-258.

Abbas, H. K., Accinelli, C., Zablotowicz, R. M., Abel, C. A., Bruns, H. A., Dong, Y., & Shier, W. T. (2008). Dynamics of mycotoxin and *Aspergillus flavus* levels in aging Bt and non-Bt corn residues under Mississippi no-till conditions. *Journal of agricultural and food chemistry*, *56*(16), 7578-7585.

Abdel-Wahhab, M. A., Nada, S. A., & Amra, H. A. (1999). Effect of aluminosilicates and bentonite on aflatoxin-induced developmental toxicity in rat. *Journal of Applied Toxicology*, *19*(3), 199-204.

Adams, T. B., Cohen, S. M., Doull, J., Feron, V. J., Goodman, J. I., Marnett, L. J., & Wagner, B. M. (2004). The FEMA GRAS assessment of cinnamyl derivatives used as flavor ingredients. *Food and chemical toxicology*, *42*(2), 157-185.

Aly, S. A., & Anwer, W. (2009). Effect of naturally contaminated feed with aflatoxins on performance of laying hens and the carryover of aflatoxin B1 residues in table eggs. *Pakistan Journal of Nutrition*, *8*(2), 181-186.

Antilla, L., & Cotty, P. J. (2001). Utilizing atoxigenic strains of *A. flavus* to manage aflatoxins in commercial cotton: Part II. Field Aspects. In J. F. Robens, J. W. Cary, & B. W. Campbell (Eds.), *Proc USDA-ARS Aflatoxin Elimination Workshop* (pp. 132-133).

Applebaum, R. S., Brackett, R. E., Wiseman, D. W., & Marth, E. H. (1982). Responses of dairy cows to dietary aflatoxin: feed intake and yield, toxin content, and quality of milk of cows treated with pure and impure aflatoxin. *Journal of Dairy Science*, *65*(8), 1503-1508.

Azzam, A. H., & Gabal, M. A. (1998). Aflatoxin and immunity in layer hens. *Avian Pathology*, *27*(6), 570-577.

Azziz-Baumgartner, E., Lindblade, K., Gieseker, K., Rogers, H. S., Kieszak, S., Njapau, H., & Rubin, C. (2005). Case-control study of an acute aflatoxicosis outbreak, Kenya, 2004. *Environmental Health Perspectives*, 1779-1783.

Bedard, L. L. & Massey, T. E. (2006) Aflatoxin B1-induced DNA damage and its repair. *Cancer Letter* 241, 174–183.

Bagherzadeh Kasmani, F., Karimi Torshizi, M. A., Allameh, A., & Shariatmadari, F. (2012). A novel aflatoxin-binding *Bacillus* probiotic: Performance, serum biochemistry, and immunological parameters in Japanese quail. *Poultry Science*, *91*(8), 1846-1853.

Bennett, R. A., Essigmann, J. M., & Wogan, G. N. (1981). Excretion of an aflatoxin-guanine adduct in the urine of aflatoxin B1-treated rats. *Cancer Research*, *41*(2), 650-654.

Blount, W. P. (1961). Turkey "X" disease. *Turkeys*, *9*(2), 52-55.

Bonna, R. J., Aulerich, R. J., Bursian, S. J., Poppenga, R. H., Braselton, W. E., & Watson, G. L. (1991). Efficacy of hydrated sodium calcium aluminosilicate and activated charcoal in reducing the toxicity of dietary aflatoxin to mink. *Archives of environmental contamination and toxicology*, *20*(3), 441-447.

Bovo, F., Corassin, C. H., Rosim, R. E., & de Oliveira, C. A. (2012). Efficiency of lactic acid bacteria strains for decontamination of aflatoxin M 1 in phosphate buffer saline solution and in skimmed milk. *Food and Bioprocess Technology*, 1-5.

Buntin, G. D., Lee, R. D., Wilson, D. M., & McPherson, R. M. (2001). Evaluation of YieldGard transgenic resistance for control of fall armyworm and corn earworm (Lepidoptera: Noctuidae) on corn. *Florida Entomologist*, 37-42.

Burt, S. (2004). Essential oils: their antibacterial properties and potential applications in foods—a review. *International journal of food microbiology*, *94*(3), 223-253.

Cary, J. W., Montalbano, B. G., & Ehrlich, K. C. (2000). Promoter elements involved in the expression of the *Aspergillus parasiticus* aflatoxin biosynthesis pathway gene avnA. *Biochimica et Biophysica Acta (BBA)-Gene Structure and Expression*, *1491*(1), 7-12.

CAST. (2003) Mycotoxins: Risks in Plant, Animal and Human Systems; No. 139. Council for Agricultural Science and Technology: Ames, IA, USA, 2003.

Celik, I., Oguz, H., Demet, O., Boydak, M., Donmez, H. H. M., Sur, E., & Nizamlioglu, F. (2000). Embryotoxicity assay of aflatoxin produced by *Aspergillus parasiticus* NRRL 2999. *British Poultry Science*, *41*(4), 401-409.

Chen, Z. Y., Brown, R. L., Damann, K. E., & Cleveland, T. E. (2010). PR10 expression in maize and its effect on host resistance against *Aspergillus flavus* infection and aflatoxin production. *Molecular plant pathology*, *11*(1), 69-81.

Chen, Z. Y., Brown, R. L., Rajasekaran, K., Damann, K. E., & Cleveland, T. E. (2006). Identification of a maize kernel pathogenesis-related protein and evidence for its involvement in resistance to *Aspergillus flavus* infection and aflatoxin production. *Phytopathology*, *96*(1), 87-95.

Cleveland, T. E., Dowd, P. F., Desjardins, A. E., Bhatnagar, D., & Cotty, P. J. (2003). United States Department of Agriculture—Agricultural Research Service research on pre-harvest prevention of mycotoxins and mycotoxigenic fungi in US crops. *Pest management science*, *59*(6-7), 629-642.

Cotty, P. J. (1989). Virulence and cultural characteristics of two *Aspergillus flavus* strains pathogenic on cotton. *Phytopathology*, *79*(7), 808-814.

Cotty, P. J. (1994). Influence of field application of an atoxigenic strain of *Aspergillus flavus* on the populations of *A. flavus* infecting cotton bolls and on the aflatoxin content of cottonseed. *Phytopathology*, *84*(11), 1270-1277.

Cotty P. J. (1997). Aflatoxin-producing potential of communities of *Aspergillus* section *Flavi* from cotton producing areas in the United States. *Mycology Research.* 101:698–704.

Cotty, P. J. (2001). Cottonseed losses and mycotoxins. *Compendium of cotton diseases. Part*, *1*, 9-13.

Cotty, P. J., & Jaime-Garcia, R. (2007). Influences of climate on aflatoxin producing fungi and aflatoxin contamination. *International Journal of Food Microbiology*, *119*(1), 109-115.

Cova, L., Wild, C. P., Mehrotra, R., Turusov, V., Shirai, T., Lambert, V., & Montesano, R. (1990). Contribution of aflatoxin B1 and hepatitis B virus infection in the induction of liver tumors in ducks. *Cancer research*, *50*(7), 2156-2163.

Cullen, J. M. & Newberne, P. M. (1993). Acute hepatotoxicity of aflatoxins. In: Eaton, D. L., Groopman, J. D. (Eds.), The Toxicology of Aflatoxins: Human Health, Veterinary, and Agricultural Significance. Academic Press, London 1–26.

de Alencar, E. R., Faroni, L. R. D. A., Soares, N. D. F. F., da Silva, W. A., & da Silva Carvalho, M. C. (2012). Efficacy of ozone as a fungicidal and detoxifying agent of aflatoxins in peanuts. *Journal of the Science of Food and Agriculture*, *92*(4), 899-905.

Dersjant-Li, Y., Verstegen, M. W., & Gerrits, W. J. (2003). The impact of low concentrations of aflatoxin, deoxynivalenol or fumonisin in diets on growing pigs and poultry. *Nutrition research reviews*, *16*(02), 223-239.

Dhanasekaran, D., Shanmugapriya, S., Thajuddin, N., & Panneerselvam, A. (2011). Aflatoxins and aflatoxicosis in human and animals. In *Aflatoxins-Biochemistry and Molecular Biology*. InTech. 12, 221-254.

Diaz, D. E. (2005) *The Mycotoxin Blue Book*, 1st ed.; Nottingham University Press: Nottingham, UK, pp. 25–56.

Dietert, R. R., Qureshi, M., Nanna, U. C., & Bloom, S. E. (1985). Embryonic exposure to aflatoxin-B1: Mutagenicity and influence on development and immunity. *Environmental mutagenesis*, *7*(5), 715-725.

Dorner, J. W. (2010). Efficacy of a biopesticide for control of aflatoxins in corn. *Journal of food protection*, *73*(3), 495-499.

Dorner, J. W., & Cole, R. J. (2002). Effect of application of nontoxigenic strains of Aspergillus flavus and A. parasiticus on subsequent aflatoxin contamination of peanuts in storage. *Journal of Stored Products Research*, *38*(4), 329-339.

Dozier, W. A., Kidd, M. T., & Corzo, A. (2008). Dietary amino acid responses of broiler chickens. *The Journal of Applied Poultry Research*, *17*(1), 157-167.

Eaton, D. L., & Gallagher, E. P. (1994). Mechanisms of aflatoxin carcinogenesis. *Annual review of pharmacology and toxicology*, *34*(1), 135-172.

Edrington, T. S., Harvey, R. B., & Kubena, L. F. (1995). Toxic effects of aflatoxin B1 and ochratoxin A, alone and in combination, on chicken embryos. *Bulletin of environmental contamination and toxicology*, *54*(3), 331-336.

Edrington, T. S., Kubena, L. F., Harvey, R. B., & Rottinghaus, G. E. (1997). Influence of a superactivated charcoal on the toxic effects of aflatoxin or T-2 toxin in growing broilers. *Poultry Science*, *76*(9), 1205-1211.

Ehrlich, K. C., & Cotty, P. J. (2004). An isolate of *Aspergillus flavus* used to reduce aflatoxin contamination in cottonseed has a defective polyketide synthase gene. *Applied Microbiology and Biotechnology*, *65*(4), 473-478.

Ehrlich, K. C., Yu, J., & Cotty, P. J. (2005). Aflatoxin biosynthesis gene clusters and flanking regions. *Journal of Applied Microbiology*, *99*(3), 518-527.

Ellis, W. O., Smith, J. P., Simpson, B. K., Oldham, J. H., & Scott, P. M. (1991). Aflatoxins in food: occurrence, biosynthesis, effects on organisms, detection, and methods of control. *Critical Reviews in Food Science & Nutrition*, *30*(4), 403-439.

El-Nezami, H., Kankaanpaa, P., Salminen, S., & Ahokas, J. (1998). Ability of dairy strains of lactic acid bacteria to bind a common food carcinogen, aflatoxin B 1. *Food and Chemical Toxicology*, *36*(4), 321-326.

Emafo P. O. (1976) Species differences in the metabolism of aflatoxin B1. African *Journal of Medical and Medicine Science* 5:55–62

Gandomi, H., Misaghi, A., Basti, A. A., Bokaei, S., Khosravi, A., Abbasifar, A., & Javan, A. J. (2009). Effect of Zataria multiflora Boiss. essential oil on growth and aflatoxin formation by Aspergillus flavus in culture media and cheese. *Food and chemical toxicology*, *47*(10), 2397-2400.

Garlich, J. D., Tung, H. T., & Hamilton, P. B. (1973). The effects of short term feeding of aflatoxin on egg production and some plasma constituents of the laying hen. *Poultry Science*, *52*(6), 2206-2211.

Giambrone, J. J., Diener, U. L., Davis, N. D., Panangala, V. S., & Hoerr, F. J. (1985). Effects of aflatoxin on young turkeys and broiler chickens. *Poultry Science*, *64*(9), 1678-1684.

Goldblatt, L. A. (1971). Control and removal of aflatoxin. *Journal of the American Oil Chemists' Society*, *48*(10), 605-610.

Gratz, S., Mykkänen, H., & El-Nezami, H. (2005). Aflatoxin B1 binding by a mixture of Lactobacillus and Propionibacterium: in vitro versus ex vivo. *Journal of food protection*, *68*(11), 2470-2474.

Guengerich, F. P., Johnson, W. W., Ueng, Y. F., Yamazaki, H., & Shimada, T. (1996). Involvement of cytochrome P450, glutathione S-transferase, and epoxide hydrolase in the metabolism of aflatoxin B1 and relevance to risk of human liver cancer. *Environmental health perspectives*, *104*(Suppl 3), 557.

Guo, B. Z., Xu, G., Cao, Y. G., Holbrook, C. C., & Lynch, R. E. (2006). Identification and characterization of phospholipase D and its association with drought susceptibilities in peanut (*Arachis hypogaea*). *Planta*, *223*(3), 512-520.

Guthrie, L. D., & Bedell, D. M. (1979). Effects of aflatoxin in corn on production and reproduction in dairy cattle. In *Proceedings, annual meeting of the United States Animal Health Association* (No. 83, p. 202).

Groopman, J. D., Cain, L. G., Kensler, T. W., & Harris, C. C. (1988). Aflatoxin exposure in human populations: measurements and relationship to cancer. *CRC Critical reviews in Toxicology*, *19*(2), 113-145.

Groopman, J. D., Jiaqi, Z., Donahue, P. R., Pikul, A., Lisheng, Z., Jun-shi, C., & Wogan, G. N. (1992). Molecular dosimetry of urinary aflatoxin-DNA adducts in people living in Guangxi Autonomous Region, People's Republic of China. *Cancer Research*, *52*(1), 45-52.

Groopman, J. D., Kensler, T. W., & Wild, C. P. (2008). Protective interventions to prevent aflatoxin-induced carcinogenesis in developing countries. *Annu. Rev. Public Health*, *29*, 187-203.

Hall, A. J., & Wild, C. P. (1994). Epidemiology of Aflatoxin. *The toxicology of aflatoxins: Human health, veterinary, and agricultural significance*.

Halver, J. E. (1969). Chapter 5: Aflatoxicosis and trout hepatoma. *Aflatoxin: Scientific background, control, and implications*, *7*, 265.

Hamid, A. S., Tesfamariam, I. G., Zhang, Y., & Zhang, Z. G. (2013). Aflatoxin B1-induced hepatocellular carcinoma in developing countries: Geographical distribution, mechanism of action and prevention (Review). *Oncology letters*, *5*(4), 1087-1092.

Haskard, C. A., El-Nezami, H. S., Kankaanpää, P. E., Salminen, S., & Ahokas, J. T. (2001). Surface binding of aflatoxin B1 by lactic acid bacteria. *Applied and environmental microbiology*, *67*(7), 3086-3091.

Hedayati, M. T., Pasqualotto, A. C., Warn, P. A., Bowyer, P., & Denning, D. W. (2007). *Aspergillus flavus*: human pathogen, allergen and mycotoxin producer. *Microbiology*, *153*(6), 1677-1692.

Havenstein, G. B., Ferket, P. R., Scheideler, S. E., & Larson, B. T. (1994). Growth, livability, and feed conversion of 1957 vs 1991 broilers when fed "typical" 1957 and 1991 broiler diets. *Poultry Science*, *73*(12), 1785-1794.

Hell, K., Fandohan, P., Bandyopadhyay, R., Kiewnick, S., Sikora, R., & Cotty, P. J. (2008). Pre-and post-harvest management of aflatoxin in maize: An African perspective. *Mycotoxins: Detection methods, management, public health and agricultural trade*, Myco-Globe, Cambridge, MA. pp. 219-229.

Henry, S. H., Bosch, F. X., Troxell, T. C., & Bolger, P. M. (1999). Reducing liver cancer--global control of aflatoxin. *Science*, *286*(5449), 2453-2454.

Hesseltine, C. W. (1979). Some important fermented foods of Mid-Asia, the Middle East, and Africa. *Journal of the American Oil Chemists' Society*, *56*(3), 367-374.

Horn, B. W. (2003). Ecology and population biology of aflatoxigenic fungi in soil. *Journal of Toxicology: Toxin Reviews*, *22*(2-3), 351-379.

Horn, B. W., & Dorner, J. W. (1998). Soil populations of *Aspergillus* species from section Flavi along a transect through peanut-growing regions of the United States. *Mycologia*, 767-776.

Horn, B. W., Moore, G. G., & Carbone, I. (2009). Sexual reproduction in *Aspergillus flavus*. *Mycologia*, *101*(3), 423-429.

Horn B. W., & Pitt J. I. (1997). Yellow mold and aflatoxin. In *Compendium of Peanut Diseases*, ed. N Kokalis-Burelle, DM Porter, R Rodriguez-Kabana, DH Smith, Subrahmanyam, pp. 44–49. St. Paul, MN: Am. Phytopathol. Soc.

Howarth, B., & Wyatt, R. D. (1976). Effect of dietary aflatoxin on fertility, hatchability, and progeny performance of broiler breeder hens. *Applied and environmental microbiology*, *31*(5), 680-684.

Hsu, I. C., Metcalf, R. A., Sun, T., Welsh, J. A., Wang, N. J., & Harris, C. C. (1991). Mutational hot spot in the p53 gene in human hepatocellular carcinomas. *Nature*, *350*(6317), 427-428.

Jiang, Y. I., Jolly, P. E., Ellis, W. O., Wang, J. S., Phillips, T. D., & Williams, J. H. (2005). Aflatoxin B1 albumin adduct levels and cellular immune status in Ghanaians. *International immunology*, *17*(6), 807-814.

Karunaratne, A., Wezenberg, E., & Bullerman, L. B. (1990). Inhibition of Mold Growth and Aflatoxin Production by *Lactobacillus* spp 1. *Journal of Food Protection*, *53*(3), 230-236.

Khan, W. A., Khan, M. Z., Khan, A., Hassan, Z. U., Rafique, S., Saleemi, M. K., & Ahad, A. (2014). Dietary vitamin E in White Leghorn layer breeder hens: a strategy to combat aflatoxin B1-induced damage. *Avian Pathology*, *43*(5), 389-395.

Klein, P. J., Buckner, R., Kelly, J., & Coulombe, R. A. (2000). Biochemical basis for the extreme sensitivity of turkeys to aflatoxin B1. *Toxicology and applied pharmacology*, *165*(1), 45-52.

Klich, M. A. (2007). *Aspergillus flavus*: the major producer of aflatoxin. *Molecular plant pathology*, *8*(6), 713-722.

Kong, Q., Shan, S., Liu, Q., Wang, X., & Yu, F. (2010). Biocontrol of Aspergillus flavus on peanut kernels by use of a strain of marine Bacillus megaterium. *International journal of food microbiology*, *139*(1), 31-35.

Kuang, S. Y., Lekawanvijit, S., Maneekarn, N., Thongsawat, S., Brodovicz, K., Nelson, K., & Groopman, J. D. (2005). Hepatitis B 1762T/1764A mutations, hepatitis C infection, and codon 249 p53 mutations in hepatocellular carcinomas from Thailand. *Cancer Epidemiology and Prevention Biomarkers*, *14*(2), 380-384.

Kubena, L. F., Harvey, R. B., Huff, W. E., Elissalde, M. H., Yersin, A. G., Phillips, T. D., & Rottinghaus, G. E. (1993). Efficacy of a hydrated sodium calcium aluminosilicate to reduce the toxicity of aflatoxin and diacetoxyscirpenol. *Poultry Science*, *72*(1), 51-59.

Lahtinen, S. J., Haskard, C. A., Ouwehand, A. C., Salminen, S. J., & Ahokas, J. T. (2004). Binding of aflatoxin B1 to cell wall components of Lactobacillus rhamnosus strain GG. *Food additives and contaminants*, *21*(2), 158-164.

LaPrade, J. C., & Manwiller, A. (1977). Relation of insect damage, vector, and hybrid reaction to aflatoxin B1 recovery from field corn. *Phytopathology*, *67*, 544-547.

Lee, J. T., Jessen, K. A., Beltran, R., Starkl, V., Schatzmayr, G., Borutova, R., & Caldwell, D. J. (2012). Mycotoxin-contaminated diets and deactivating compound in laying hens: 1. Effects on performance characteristics and relative organ weight. *Poultry science*, *91*(9), 2089-2095.

Leeson, S. (1995). Aflatoxins "Poutry metabolic disorders and mycotoxins" Leeson, S., Diaz, G. & Summers, JD (Eds.). pp: 248-279.

Ledoux, D. R., Rottinghaus, G. E., Bermudez, A. J., & Alonso-Debolt, M. (1999). Efficacy of a hydrated sodium calcium aluminosilicate to ameliorate the toxic effects of aflatoxin in broiler chicks. *Poultry Science*, *78*(2), 204-210.

Lillehoj, E. B., Ciegler, A., & Hall, H. H. (1967). Aflatoxin B1 uptake by *Flavobacterium aurantiacum* and resulting toxic effects. *Journal of bacteriology*, *93*(1), 464-471.

Link, H. F. (1809). *Observationes in ordines plantarum naturales: Dissertatio Ima complectens anandrarum ordines epiphytas, mucedines gastromycos et fungos*. [Plants Natural observations are in order: The lowest portion comprising anandrum orders epiphytes, Mucedinei gastromycos and mushrooms]

Lussenhop, J., & Wicklow, D. T. (1990). Nitidulid beetles (Nitidulidae: Coleoptera) as vectors of *Aspergillus flavus* in pre-harvest maize. *Transactions of the Mycological Society of Japan*, *31*, 63-74.

Lynch, R. E., Guo, B., Timper, P., & Wilson, J. P. (2003). United States Department of Agriculture–Agricultural Research Service research on improving host-plant resistance to pests. *Pest management science*, *59*(6-7), 718-727.

Mayer, Z., Bagnara, A., Färber, P., & Geisen, R. (2003). Quantification of the copy number of *nor1*, a gene of the aflatoxin biosynthetic pathway by real-time PCR, and its correlation to the cfu of Aspergillus flavus in foods. *International Journal of Food Microbiology*, *82*(2), 143-151.

Mejía-Teniente, L., Chapa-Oliver, A. M., Vazquez-Cruz, M. A., Torres-Pacheco, I., & Guevara-González, R. G. (2011). Aflatoxins Biochemistry and Molecular Biology-Biotechnological Approaches for Control in Crops. In *Aflatoxins-Detection, Measurement and Control*. InTech.16, 317-354.

Miazzo, R., Rosa, C. A. R., Carvalho, E. D. Q., Magnoli, C., Chiacchiera, S. M., Palacio, G., & Dalcero, A. (2000). Efficacy of synthetic zeolite to reduce the toxicity of aflatoxin in broiler chicks. *Poultry Science*, *79*(1), 1-6.

Micco, C., Miraglia, M., Onori, R., Brera, C., Mantovani, A. L., Ioppolo, A., & Stasolla, D. (1988). Long-term administration of low doses of mycotoxins to poultry. 1. Residues of aflatoxin B1 and its metabolites in broilers and laying hens. *Food Additives & Contaminants*, *5*(3), 303-308.

Mishra, H. N., & Das, C. (2003). A review on biological control and metabolism of aflatoxin. 43(3), 245-264.

Moss, M. O. (1998). Recent studies of mycotoxins. *Journal of Applied Microbiology*, *84*(s1), 62S.

Newberne, P. M., & Butler, W. H. (1969). Acute and chronic effects of aflatoxin on the liver of domestic and laboratory animals: a review. *Cancer research*, *29*(1), 236-250.

Oliveira IV, C. A. F., Kobashigawa, E., Reis, T. A., Mestieri, L., Albuquerque, R., & Correa, L. M. B. (2000). Aflatoxin B1 residues in eggs of laying hens fed a diet containing different levels of the mycotoxin. *Food Additives and Contaminants*, *17*(6), 459-462.

Oliveira, C. A., Rosmaninho, J. F., Butkeraitis, P., Correa, B., Reis, T. A., Guerra, J. L., & Moro, M. E. (2002). Effect of low levels of dietary aflatoxin B1 on laying Japanese quail. *Poultry science*, *81*(7), 976-980.

Oliveira, C. F., Rosmaninho, J. F., Castro, A. L., Butkeraitis, P., Reis, T. A., & Corrêa, B. (2003). Aflatoxin residues in eggs of laying Japanese quail after long-term administration of rations containing low levels of aflatoxin B1. *Food Additives and Contaminants*, *20*(7), 648-653.

Ortatatli, M., & Oğuz, H. (2001). Ameliorative effects of dietary clinoptilolite on pathological changes in broiler chickens during aflatoxicosis. *Research in Veterinary Science*, *71*(1), 59-66.

Oznurlu, Y., Celik, I., Sur, E., Ozaydın, T., Oğuz, H., & Altunbaş, K. (2012). Determination of the effects of aflatoxin B1 given in ovo on the proximal tibial growth plate of broiler chickens: histological, histometric and immunohistochemical findings. *Avian pathology*, *41*(5), 469-477.

Pandey, I., & Chauhan, S. S. (2007). Studies on production performance and toxin residues in tissues and eggs of layer chickens fed on diets with various concentrations of aflatoxin AFB1. *British poultry science*, *48*(6), 713-723.

Payne, G. A. (1998). Process of contamination by aflatoxin-producing fungi and their impact on crops. *Mycotoxins in agriculture and food safety*, *9*, 279-306.

Peltonen, K. D., El-Nezami, H. S., Salminen, S. J., & Ahokas, J. T. (2000). Binding of aflatoxin B1 by probiotic bacteria. *Journal of the Science of Food and Agriculture*, *80*(13), 1942-1945.

Phillips, T. D. (1999). Dietary clay in the chemoprevention of aflatoxin-induced disease. *Toxicological Sciences*, *52*(suppl 1), 118-126.

Pier, A. C. (1992). Major biological consequences of aflatoxicosis in animal production. *Journal of Animal Science*, *70*(12), 3964-3967.

Pitt, J. I., & Hocking, A. D. (2006). Mycotoxins in Australia: biocontrol of aflatoxin in peanuts. *Mycopathologia*, *162*(3), 233-243.

Pittet, A. (1998). Natural occurrence of mycotoxins in foods and feeds: An update review. *Revue de Medecine Veterinaire (France)*. http://agris.fao.org/agris-search/search.do?recordID=FR1998004063.

Qian, G. S., Ross, R. K., Yu, M. C., Yuan, J. M., Gao, Y. T., Henderson, B. E., & Groopman, J. D. (1994). A follow-up study of urinary markers of aflatoxin exposure and liver cancer risk in Shanghai, People's Republic of China. *Cancer Epidemiology and Prevention Biomarkers*, *3*(1), 3-10.

Quezada, T., Cuellar, H., Jaramillo-Juarez, F., Valdivia, A. G., & Reyes, J. L. (2000). Effects of aflatoxin B 1 on the liver and kidney of broiler chickens during development. *Comparative Biochemistry and Physiology Part C: Pharmacology, Toxicology and Endocrinology*, *125*(3), 265-272.

Qureshi, M. A., Brake, J., Hamilton, P. B., Hagler, W. M., & Nesheim, S. (1998). Dietary exposure of broiler breeders to aflatoxin results in immune dysfunction in progeny chicks. *Poultry Science*, *77*(6), 812-819.

Razzaghi-Abyaneh, M., Shams-Ghahfarokhi, M., Yoshinari, T., Rezaee, M. B., Jaimand, K., Nagasawa, H., & Sakuda, S. (2008). Inhibitory effects of Satureja hortensis L. essential oil on growth and aflatoxin production by Aspergillus parasiticus. *International journal of food microbiology*, *123*(3), 228-233.

Rawal, S., Kim, J. E., & Coulombe, R. (2010). Aflatoxin B 1 in poultry: toxicology, metabolism and prevention. *Research in veterinary science*, *89*(3), 325-331.

Rezaei, M., Mohammadpourfard, I., Malekirad, A. A., Ghasemikhah, R., & Sayadi, M. (2014). Aflatoxin contamination of feed materials in Qom province, Iran. *Iranian Journal of Toxicology*, *8*(24), 1025-1027.

Robens, J., & Cardwell, K. (2003). The costs of mycotoxin management to the USA: management of aflatoxins in the United States. *Journal of Toxicology: Toxin Reviews*, *22*(2-3), 139-152.

Ross, R. K., Yu, M. C., Henderson, B. E., Yuan, J. M., Qian, G. S., Tu, J. T., & Groopman, J. D. (1992). Urinary aflatoxin biomarkers and risk of hepatocellular carcinoma. *The Lancet*, *339*(8799), 943-946.

Rubens, J., & Cardwell, K. F. (2005). The cost of mycotoxin management in the United States. *Aflatoxin and Food Safety*, 1-12.

Salamci, E., Kordali, S., Kotan, R., Cakir, A., & Kaya, Y. (2007). Chemical compositions, antimicrobial and herbicidal effects of essential oils isolated from Turkish *Tanacetum aucheranum* and *Tanacetum chiliophyllum* var. *chiliophyllum*. *Biochemical systematics and ecology*, *35*(9), 569-581.

Samarajeewa, U., Sen, A. C., Cohen, M. D., & Wei, C. I. (1990). Detoxification of Aflatoxins in Foods and Feeds by Physical and Chemical Methods 1. *Journal of food protection*, *53*(6), 489-501.

Scheideler, S. E. (1993). Effects of various types of aluminosilicates and aflatoxin B1 on aflatoxin toxicity, chick performance, and mineral status. *Poultry Science*, *72*(2), 282-288.

Shah, N., & Wu, X. (1999). Aflatoxin B1 binding abilities of probiotic bacteria. *Bioscience and microflora*, *18*(1), 43-48.

Shetty, P. H., & Jespersen, L. (2006). Saccharomyces cerevisiae and lactic acid bacteria as potential mycotoxin decontaminating agents. *Trends in Food Science & Technology*, *17*(2), 48-55.

Sims, W. M., Kelley, D. C., & Sanford, P. E. (1970). A study of aflatoxicosis in laying hens. *Poultry science*, *49*(4), 1082-1084.

Sinha, K. K., Sinha, A. K., & Prasad, G. (1993). The effect of clove and cinnamon oils on growth of and aflatoxin production by *Aspergillus flavus*. *Letters in Applied Microbiology*, *16*(3), 114-117.

Sur, E., & Celik, I. (2003). Effects of aflatoxin B1 on the development of the bursa of Fabricius and blood lymphocyte acid phosphatase of the chicken. *British poultry science*, *44*(4), 558-566.

Teniola, O. D., Addo, P. A., Brost, I. M., Färber, P., Jany, K. D., Alberts, J. F., & Holzapfel, W. H. (2005). Degradation of aflatoxin B 1 by cell-free extracts of Rhodococcus erythropolis and Mycobacterium fluoranthenivorans sp. nov. DSM44556 T. *International journal of food microbiology*, *105*(2), 111-117.
Teniola, O. D., Addo, P. A., Brost, I. M., Färber, P., Jany, K. D., Alberts, J. F., & Holzapfel, W. H. (2005). Degradation of aflatoxin B 1 by cell-free extracts of Rhodococcus erythropolis and Mycobacterium fluoranthenivorans sp. nov. DSM44556 T. *International journal of food microbiology*, *105*(2), 111-117.
Van Egmond, H. P., & Jonker, M. A. (2004). Worldwide regulations on aflatoxins—the situation in 2002. *Journal of Toxicology: Toxin Reviews*, *23*(2-3), 273-293.

Van der Zijden, A. S. M., Koelensmid, W. B., Boldixgh, J., Barrett, C. B., Ord, W. O., & Philp, J. (1962). Isolation in crystalline form of a toxin responsible for Turkey X disease. *Nature, 195*(4846).

Vardon, P., McLaughlin, C., & Nardinelli, C. (2003). Potential economic costs of mycotoxins in the United States. *Council for Agricultural Science and Technology Task Force Report*, pp.139.

Verma, J., Johri, T. S., Swain, B. K., & Ameena, S. (2004). Effect of graded levels of aflatoxin, ochratoxin and their combinations on the performance and immune response of broilers. *British poultry science*, *45*(4), 512-518.

Westlake, K., Mackie, R. I., & Dutton, M. F. (1989). In vitro metabolism of mycotoxins by bacterial, protozoal and ovine ruminal fluid preparations. *Animal Feed Science and Technology*, *25*(1-2), 169-178.

Whitlow, L. W., & Hagler, W. M. (2010). Mold and mycotoxin issues in dairy cattle: effects, prevention and treatment. *Advance Dairy Technology*, *20*, 195-209.

Wicklow, D. T., & Donahue, J. E. (1984). Sporogenic germination of sclerotia in *Aspergillus flavus* and *Aspergillus parasiticus*. *Transactions of the British Mycological Society*, *82*(4), 621-624.

Wicklow, D. T., Wilson, D. M., & Nelsen, T. C. (1993). Survival of *Aspergillus flavus* sclerotia and conidia buried in soil in Illinois or Georgia. *Phytopathology-New York and Baltimore then st Paul*, *83*, 1141-1141.

Wild, C. P., Hudson, G. J., Sabbioni, G., Chapot, B., Hall, A. J., Wogan, G. N., & Groopman, J. D. (1992). Dietary intake of aflatoxins and the level of albumin-bound aflatoxin in peripheral blood in The Gambia, West Africa. *Cancer Epidemiology and Prevention Biomarkers*, *1*(3), 229-234.

Williams, J. H., Phillips, T. D., Jolly, P. E., Stiles, J. K., Jolly, C. M., & Aggarwal, D. (2004). Human aflatoxicosis in developing countries: a review of toxicology, exposure, potential health consequences, and interventions. *The American journal of clinical nutrition*, *80*(5), 1106-1122.

Yarru, L. P., Settivari, R. S., Gowda, N. K. S., Antoniou, E., Ledoux, D. R., & Rottinghaus, G. E. (2009). Effects of turmeric (*Curcuma longa*) on the expression of hepatic genes associated with biotransformation, antioxidant, and immune systems in broiler chicks fed aflatoxin. *Poultry science*, *88*(12), 2620-2627.

Yiannikouris, A., & Jouany, J. P. (2002). Mycotoxins in feeds and their fate in animals: a review. *Animal Research*, *51*(2), 81-99.

Yin, H. B., Chen, C. H., Kollanoor-Johny, A., Darre, M. J., & Venkitanarayanan, K. (2015a). Controlling *Aspergillus flavus* and *Aspergillus parasiticus* growth and aflatoxin production in poultry feed using carvacrol and *trans*-cinnamaldehyde. *Poultry science*, *94*(9), 2183-2190.

Yin, H. B., Chen, C. H., Upadhyaya, I., Darre, M., Donoghue, A., Donoghue, D., & Venkitanarayanan, K. (2015b). Phytochemicals reduce aflatoxin-induced toxicity in chicken embryos. *Poultry Science* 94, (E-Supplement 1), 141-142.

Yin, H., C. Chen, I. Upadhyaya, A. Upadhyay, S. Fancher, J. Li, M. S. Nair, S. Mooyottu, M. I. Khan, M. J. Darre, A. M. Donoghue, D. J. Donoghue, & Venkitanarayanan K. (2015c). Efficacy of in-feed supplementation of plant-derived antimicrobials in reducing aflatoxicosis in chickens. Poultry Science 94, (E-Supplement 1), 9.

Yu, J., Cleveland, T. E., Nierman, W. C., & Bennett, J. W. (2005). *Aspergillus flavus* genomics: gateway to human and animal health, food safety, and crop resistance to diseases. *Revista iberoamericana de micología*, *22*(4), 194-202.

Yu, J., Whitelaw, C. A., Nierman, W. C., Bhatnagar, D., & Cleveland, T. E. (2004). *Aspergillus flavus* expressed sequence tags for identification of genes with putative roles in aflatoxin contamination of crops. *FEMS Microbiology letters*, *237*(2), 333-340.

Yunus, A. W., Razzazi-Fazeli, E., & Bohm, J. (2011). Aflatoxin B1 in affecting broiler's performance, immunity, and gastrointestinal tract: A review of history and contemporary issues. *Toxins*, *3*(6), 566-590.

Zakharian, R. A., Israelian, I., Agabalian, A. S., Tatevosian, P. E., & Akopian, S. M. (1978). *Bacillus thuringiensis* plasmid DNA. *Mikrobiologiia*, *48*(2), 226-229.

Zhu, J. Q., Zhang, L. S., Hu, X., Xiao, Y., Chen, J. S., Xu, Y. C., & Chu, F. S. (1987). Correlation of dictary aflatoxin B1 levels with excretion of aflatoxin M1 in human urine. *Cancer Research*, *47*(7), 1848-1852.

In: Focus on Aflatoxins Research
Editors: P. Jan and S. Sylvester
ISBN: 978-1-53612-569-6

Chapter 4

MYCOBIOTA AND AFLATOXINS IN FEED FOR GOATS AND THEIR RESIDUES IN MILK

Edlayne Gonçalez[1,*], PhD, Janaina Lara da Silva[2], MD, Tatiana Emy de Freitas[3], MD, Christiane Ceriani Aparecido[4], PhD, Daiane Hansen[1], PhD, and Joana D'arc Felicio[1], PhD

[1]Centro de Pesquisa e Desenvolvimento de Sanidade Animal, Instituto Biológico, São Paulo, Brazil

[2]Colégio Dom Henrique, São Paulo, SP, Brazil

[3]Instituto de Ciências Biomédicas, Universidade de São Paulo, São Paulo, Brazil

[4]Centro de Pesquisa e Desenvolvimento de Sanidade Vegetal, Instituto Biológico, São Paulo, Brazil

---

[*] Corresponding author: goncalez@biologico.sp.gov.br.

ABSTRACT

Aflatoxins are highly toxic and carcinogenic metabolites produced by *Aspergillus* species in food, feed and agricultural commodities. Several studies on aflatoxins in feed and bovine milk have been reported, but in feed and milk the goats are scarce. This study aimed to evaluate the presence of aflatoxins in diets intended for dairy goats and their residues in milk and identify the mycobiota present in the feed and forage. *Aspergillus flavus* and *A. parasiticus* strains were isolated and characterize based on a morphological and molecular characterization and mycotoxigenic ability. The goat dairy diets were collected monthly from 10 goat milk farms, totaling 128, 106 and 120 samples of concentrate, forage and milk, respectively. For the isolation of the mycobiota the surface plating method was used. *Aspergillus*, *Penicillium*, and *Fusarium* were the main fungi producing mycotoxins isolated. The morphological and molecular characterization and mycotoxigenic ability were used for *A. flavus* and *A. parasiticus* identification. The *Aspergillus* spp. from feed 39% produced aflatoxins B_1 and B_2, and 2.0% of the strains produced aflatoxins B_1, B_2, G_1, and G_2. The strains from forage were producers of aflatoxins B_1 and B_2 (37%), whereas 49% have shown no toxigenic ability. The *aflD* and *aflR* genes were used by PCR and PCR-RFLP, respectively. Aflatoxins have been extracted and purified in an immunoaffinity column specific for aflatoxins B_1, B_2, G_1, G_2 and M_1. The analyses for detection and quantification of toxins were performed by high-performance liquid chromatography. From the analyses of the feeds, 2.34% of these were contaminated with aflatoxins, but aflatoxin M_1 was not detected in the milk samples. Despite low contamination of feed, the presence of toxigenic species in samples for lactating goats indicates a potential risk of contamination of dairy products, if they are exposed to environmental conditions favorable to fungal growth and mycotoxin production.

1. INTRODUCTION

The mycotoxigenic fungi involved in the food and feed chain belong mainly to the *Aspergillus, Penicillium, Fusarium* and *Alternaria* genera. Aspergillus section Flavi includes three species, *A. flavus, A. parasiticus* and *A. nomius,* producers of aflatoxins, highly toxic compounds of concern in food safety. *A. flavus* also produces other mycotoxin such as cyclopiazonic acid (CPA) [1].

The identification of Aspergillus Section Flavi has been traditionally based on morphological and biochemical characterization. Conidial wall ornamentation is regarded as the primary morphological diagnostic character for separation of *A. flavus* and *A. parasiticus*. The mycotoxigenic profile (regarding aflatoxins B and G and CPA) of these strains has been routinely used for identification [2]. Molecular techniques have been used to differentiate these two species [2, 3]. However, it is to be emphasized that the PCR detection of *A. flavus, A. parasiticus* or *A. nomius* is no guarantee of aflatoxin production since gene other than those involved in the biosynthesis of aflatoxins are not target for amplification [4].

Many agricultural products are invaded by fungi before, during and after harvest; at drying, transport and/or storage. The interaction between water activity and temperature is a most critical determinant for fungal growth and mycotoxin production [5].

The complex diet of ruminants, consisting of forages, concentrates, and silages, can be a source of diverse mixture of mycotoxins that contaminate individual feed components [6]. Both, feed grains (e.g., corn, wheat, barley, cottonseed) and forages support fungal growth that leads to mycotoxin production [7, 8].

Diseases caused by mycotoxins are characterized by diffuse syndromes, however, presenting injuries in certain organs like liver, kidney, epithelial tissue and central nervous system, depending on the kind of toxin. It is also possible the simultaneous occurrence of two or more mycotoxins, which may lead to the potentiation of the toxic effects over the susceptible organism [9].

In general, peanuts, corn, cotton seed are among the grains with a higher risk of contamination [10], and are often invaded before harvest by *A. flavus*, producer of contamination of animal products, once the mycotoxins and their metabolites can be excreted in milk, meat, and eggs [11, 12].

Aflatoxin B_1 (AFB_1) ingested by cattle, goats or lactating humans and can be biotransformed in the liver in aflatoxin M_1 (AFM_1), and hence excreted in milk [12]. AFM_1 is the hydroxylated metabolite of AFB_1, formed by cytochrome p450-associated enzymes in liver and secreted into milk through the mammary gland of lactating animals. AFM_1 has been classified in Group I as possibly carcinogenic for humans by the International Agency for Research on Cancer (IARC) [13]. When an animal eats food contaminated with 0.5% of AFB_1, 5.0% of this toxin ingested is biotransformed to AFM_1 [14]; nevertheless, in goats this percentage is from 0.18% to 3.0% [15].

AFM_1 is stable at high temperature. Several studies have investigated the distribution/stability of AFM_1 from milk to milk products [16], and then are not degraded by various thermal processing such as pasteurization and ultra-high-temperature (UHT) treatment [17].

Although most references in the literature report the presence of AFM_1 in cow milk [11, 18-20], it is known that this toxin can easily occur in milk of sheep, goats, buffaloes and camels [18, 21], but there are few reports about the presence of AFM_1 in goat milk [22]. Flores-Flores (2015) [23] related on the AFM_1 incidence in sheep and goat's milk in Europe and Asia has indicated that it might be above 30%, although very few samples had a level that exceeded the maximum permissible level established by European Commission. Baharami (2016) [24] related AFM_1 contamination in cow, goat and sheep raw milks, cheese and yoghurt samples. Co-occurrence between AFM_1 and CPA in bovine milk has been the subject of studies by other authors [12, 20], however there are no reports about CPA in goat milk.

Goat production is an important part of the economy to several countries. García (2015) [25] related that the world's goat population increased by around 55% between 1991 and 2011 and the production of goat milk increased by around 70% between 1991 and 2011. The increasing consumption of goat milk has been observe in the world due it is less allergenic [26] and higher digestibility for human [27].

In Brazil, the State of São Paulo has environmental factors conducive to the consolidation of goat and sheep production, and the scientific and technological advances contribute to increased rates of production and productivity of animals. Since the aflatoxins and CPA can be toxic compounds of concern in milk, it is very important to offer a quality feed for dairy goats. The ovine and caprine culture has grown in recent years in the State of São Paulo, being the southwestern of São Paulo is the largest producer of goat's milk. Therefore, to check the co-occurrence of AFM_1 and CPA in goat milk and its derivatives deserves much attention due to their toxic effects on humans and animals.

Figure 1. Property of goat milk located in southwestern São Paulo, Ibiuna city.

So, the aim of this work was to identify the mycobiota present in the feed and forage for dairy goat and isolate and characterize the *A. flavus* and *A. parasiticus* strains based on a polyphasic approach involving morphological and molecular characterization, and mycotoxigenic ability, and monitor the levels of aflatoxins B_1, B_2, G_1 and G_2, and CPA in feed intended for dairy goats and their residues in bulk milk from familiar properties located in southwestern São Paulo (Figure 1).

2. Material and Methods

2.1. Samples

Samples of dairy goat feed, forage and milk were collected in 10 producing properties of goat milk located in southwestern São Paulo: 3 in Ibiúna (P1, P2, and P3), 2 in Piedade (P4 and P5), 1 in Alambari (P6), 1 in Capão Bonito (P7), 1 in Guareí (P8), and 2 in Porto Feliz (P9 and P10). The breed of goats was Saneen (P1, P2, P3, P4, P7, P8 and P9), Anglo Nubian (P10) and Alpine (P5, P6 and P7). The samples were collected monthly from May 2010 to September 2011, totalizing 128 samples of feed, 106 of forage and 120 samples of milk. In all properties, the feeds are prepared using corn, cottonseed, soy beans and salt and they were stored in barrels, each of them with 20 kg capacity. Feed and forage samples (500 g) were collected from nine different points of each barrel: three points at the upper third, three points at the medium third, and three points at the lower third (Sassahara et al., 2003). The samples were homogenized and stored in plastic containers, and sent to the laboratory for analysis of mycotoxins (aflatoxins and cyclopiazonic acid) and water activity. The bulk milk samples (250 mL each) were collected in duplicate and conditioned in sterile 500 mL bottles on the day of the feed collection. The samples were frozen until the time of mycotoxin analysis.

2.2. Water Activity Determination

The water activity (Aw) of goat feed and forage samples was determined by automatic analysis using Aqualab 4TE (Decagon Devices Inc., Pullman, WA).

2.3. Aflatoxins Analysis in Feed for Goats

For analysis of aflatoxins B_1 (AFB_1), B_2 (AFB_2), G_1 (AFG_1) and G_2 (AFG_2) in feed, 50 g of each sample, after addition of 5.0 g of NaCl, was transferred to a wide-mouth flask; next, addition of 100 mL of methanol: water (80:20 v/v). After stirring for 30 minutes with a horizontal mechanical stirrer, the flask content was filtered through filter paper, and 10 mL of the extract was diluted in 40 mL of distilled water. Then, 5.0 mL of the filtered extract (5.0 mL = 0.5 g of the sample) was transferred to an immunoaffinity column with a flow rate of ca. 1-2 drops/s until allow the air to pass through the column. After completion of the operation, the column was washed with 10 mL of distilled water with a flow rate of ca. 2 drops/s and eluted twice with 1.0 mL of HPLC grade methanol with a flow rate of ca. 1-2 drops/s. After elution, the extract was collected and stored at freezing temperature until the measurement time. The high-performance liquid chromatography (HPLC) used for analysis of mycotoxins was from Shimadzu model LC 6AD, equipped with fluorescence detector RF-551 Shimadzu and ultraviolet (UV) detector - Shimazu SPD-6AV. Aflatoxins were identified and quantified at the following chromatographic conditions: as mobile phase, methanol: acetonitrile: 0.1% phosphoric acid (24:24:52 v/v/v); flow rate of 1.0 mL/min; fluorescence detector (at 364 nm excitation and 440 nm emission), ODS C18 column of 4.6 × 250 mm, 5 μm and oven temperature of 30°C [28]. The quantification method of the samples was via external standard calibration curve using 5 points for each aflatoxin. The points were: 8.25, 11.30, 16.50, 24.75, and 33.0 ng/mL for AFB_1; 1.01, 1.4, 2.02, 3.03, and 4.04 ng/mL for AFB_2; 2.89, 3.55, 5.79, 8.68 and 11.50 ng/mL for AFG_1; and 1.92, 2.67, 3.84, 5.76, 7.68 ng/mL for AFG_2. The squared correlation coefficient (r^2) was 0.991 for AFB_1, 0.998 for AFB2, 0.964 for AFG_1 and 0.999 for AFG_2. The quantification limits were established for AFB_1, AFB_2, AFG_1 and AFG_2 at 3.5 ng/g, 0.1 ng/g, 2.0 ng/g, and 0.15 ng/g, respectively. The detection limits were 2.0 ng/g, 0.05 ng/g, 1.0 ng/g, and 0.05 ng/g for AFB_1, AFB_2, AFG_1 and AFG_2, respectively. The identity of the aflatoxins was confirmed in accordance with Tarin (2004) [29], where 500 mL of each sample was dried under nitrogen, diluted with 50 mL of trifluoroacetic acid (TFA) and 200 mL of hexane at 40ºC. The solution was stirred for 15 minutes, dried under nitrogen and redissolved in 200 mL of acetonitrile: water (3:7 v/v) and analyzed by high performance liquid chromatography (HPLC).

2.4. Aflatoxin M_1 Analyze in Goat's Milk

Aflatoxin M_1 was analyzed in the milk samples according to the methodology described by Dragacci (2001) [30], where 50 mL of each milk sample was centrifuged to remove fat. The resulting skimmed milk was applied to an immunoaffinity column with a flow rate of 1 to 2 drops/s and, afterwards, the column was washed with distilled water. The sample was eluted with 1.25 mL of acetonitrile: methanol (3:2 v/v) and subsequently collected to be washed with 1.25 mL of water and again collected in the same vial. Twenty microliters (20 μL) of the extracted sample was injected into the high performance liquid chromatography column, with the mobile phase water:acetonitrile: methanol (68:24:8) at a flow rate of 1.0 mL/min and a fluorescence detector at 360 nm excitation and 440 nm emission [30]. For quantification of the samples, a 5-points calibration curve (0.355, 0.71, 0.177, 1.704, and 2.148 μg/mL) was used, and the squared correlation coefficient was 0.995. The quantification and detection limits were 0.09 ng/mL and 0.07 ng/mL, respectively.

2.5. CPA Extraction in Feed for Goats

CPA was analyzed in samples of feed, where 25 g of each sample was transferred to a wide-mouth flask plus 100 mL of methanol: 2.0% sodium bicarbonate in water (7:3 v/v). After stirring for 30 minutes with a horizontal mechanical stirrer, the flask content was filtered through filter paper. An aliquot of extract (50 mL) was transferred to a separatory funnel, followed by addition of 100 mL of hexane. After a gentle stirring, the hexane fraction was discarded, followed by addition of 50 mL of a 10%

KCl aqueous solution. This solution was acidified with 2.0 mL of 6 N HCl. CPA was eluted with 50 mL of chloroform twice. After adding anhydrous sodium sulfate, the chloroform fraction was filtered and evaporated to dryness in a rotary evaporator. The residue was resuspended in 10 mL of chloroform and transferred to a solid phase extraction column Micotox® MS2300 (Micotox Ltd., DC Bogota, Colombia) at a flow rate of 1-2 drops/s until allow the air to pass through the column. After completion of the operation, the column was washed with 10 mL of ethyl ether, 10 mL of chloroform: acetone (1:1 v/v) and 10 mL of chloroform: methanol (95:5 v/v). CPA was eluted with 10 mL of chloroform: methanol (75:25 v/v) at a flow rate of 2.0 mL/min. The extract was evaporated under nitrogen to dryness and diluted in 1.0 mL of HPLC grade methanol and stored at freezing temperature until the measurement time.

2.6. CPA Extraction in Goat's Milk

Concerning milk, CPA was extracted using 25 mL of each sample with 25 mL methanol: 2.0% sodium bicarbonate in water (7:3 v/v) and 100 mL of hexane. After 3 minutes of centrifugation, the hexane fraction was then discarded, and the sample acidified with 2.0 mL of 6 N HCl and transferred to a separatory funnel. CPA was eluted with 100 mL of chloroform twice. After adding anhydrous sodium sulfate, the chloroform fraction was filtered and evaporated to dryness in a rotary evaporator. The residue was taken up in 5.0 mL of chloroform and transferred to a solid phase extraction column Micotox® MS2300 (Micotox Ltd., DC Bogota, Colombia) at a flow rate of about 1-2 drops/s until allow the air to pass through the column. After completion of the operation, the column was washed with 10 mL of ethyl ether, 5.0 mL of chloroform: acetone (1:1 v/v). CPA was eluted with 10 ml of chloroform: methanol (75:25 v/v) at a flow rate of 2.0 mL/min. The extract was evaporated under nitrogen to dryness and diluted in 1.0 mL of HPLC grade methanol. The extract was stored at freezing temperature until measurement time [31].

2.7. CPA Analysis by HPLC

Cyclopiazonic acid was identified by high performance liquid chromatography using as a mobile phase acetonitrile: ammonium acetate buffer (0.05 M, pH 5) (8:2 v/v), at a flow rate 0.6 mL/min, UV detector 284 nm, ODS C8 column of 4.6 × 250 mm and 5.0 μm. The quantification method of the samples was via external calibration by a 5-points calibration curve (0.765, 1.531, 3.062, 6.125, 12.247 g/mL), and the squared correlation coefficient was 0.995. The limits of detection and quantification were 0.03 μg/g and 0.005 μg/g, respectively.

2.8. Identification and Enumeration of the Mycobiota from Goat Feed and Forage

The animal feed and forage samples were apportioned in 10 g aliquots and homogenized for 30 minutes in bottles containing 90 mL of sterile distilled water. Aliquots with 0.1 mL of the dilutions in serial from 10-1 to 10-6 of the samples were plates in duplicate using the surface methods in potato agar dextrose (PDA) medium. The plates were incubated at 25°C for 5 days, but the observations were made daily. The colonies were identified at the genus level, and those belonging to the genus *Aspergillus* were identified at the species level according to [32].

2.9. Molecular Characterization of *A. flavus* and *A. parasiticus*

2.9.1. DNA Extration

The strains characterized as the genus *Aspergillus* (233) were submitted to molecular identification. The strains of *A. flavus* and *A. parasiticus* were maintained in a tube containing PDA at 25°C at 7 days, and a loop full of spores from each strain was transferred to 1.5 mL of lysis buffer with 1.0 g of sterile acid-washed 0.4 to 0.6 mm diameter glass beads and vortexed for 5.0 min at maximum speed. Proteins and polysaccharides were precipitated by 750 μL of cold 3.0 M sodium acetate, pH 5.5. The solution was mixed, placed at -20°C for 10 min and centrifuged at 5000 g for 10 min (4°C) for twice. The supernatant was precipitated with one volume of cold isopropanol. The solution was mixed gently, incubated for 1.0 h at -20°C and centrifuged at 5000 g for 10 min at 4°C. DNA pellet was washed twice with 1.0 mL of cold 70% ethanol, centrifuged at 5000 g for 5.0 min at 4oC. DNA was diluted in 100 μL of ultrapure water and stored at -20°C [2]. The DNA was quantified by measuring the absorption at 260nm in a Hitachi U-2000 spectrophotometer.

2.9.2. PCR Amplification of aflD Gene

The gene *aflD* was tested for all strains and the *aflD* primers were specifically designed according to the described by Rodrigues (2009) [2]. The PCR amplifications were performed using about 1μg of fungal template DNA, 200 pmol of each primer (Nor1Forward: 5′-ACC GCT ACG CCG GCA CTC TCG GCA C-3′ and Nor2-Reverse: 5′-GTT GGC CGC CAG CTT CGA CAC TCC G-3′), MgC12-free reaction buffer, 2.0 mM MgC12, 2.5 U of Taq polymerase and 0.2 mM of each dNTP. PCR was carried out under the following conditions: one cycle at 94°C for 3.0 min; 30 cycles at 94°C for 1.0 min, at 55°C for 1.0 min and at 72°C for 1.0 min; and at 72°C for 10 min in a final extension.

2.9.3. Restriction Site Analysis of PCR Products

After analysis of PCR products by amplification of *aflD* gene, only the samples positive for *Aspergillus* fungi were used in the restriction site analysis. First, a new PCR was used to amplify two target fragments on *A. flavus* and *A. parasiticus*, and the *aflR* primer sequences were designed according to the described by Somashekar (2004) [3] to amplify a fragment of 796 pb (*aflR*1-Forward: 5′-AAC CGC ATC CAC AAT CTC AT-3′ and *aflR*2-Reverse: 5′-AGT GCA GTT CGC TCA GAA CA-3′). The reaction mixture consisted of 1.0 μg of fungal template DNA, 50 pmol of each primer, $MgCl_2$-free reaction buffer, 2.0 mM $MgCl_2$, 0.5 U of Taq polymerase and 0.2 mM of each dNTP. PCR was carried out under the following conditions: one cycle at 94°C for 5.0 min; 30 cycles at 94°C for

3min, at 50ºC for 1.25 min and at 72ºC for 1.40 min; and at 72ºC for 10 min in a final extension. Then, the PCR products were digested with *Hinc*II to analyze the restriction site. Each restriction reaction contained 10 µL of PCR product (about 50 µg DNA), 10 units of *Hinc*II enzyme and 2.0 µL of 10 x digestion buffer as recommended by the manufacturer. The reaction mixture was incubated at 37ºC for 4.0 h. The sample was frozen and dried in nitrogen gas. After, the precipitate was dissolved in 10 µL of water, and the resulting fragments were separated by electrophoresis on a 1.5% agarose gel. The primers were synthetized by Bioneer Corporation, South Korea, and all other reagents used in PCR and restriction analysis were purchased by Fermentas, Lithuania, part of Thermo Fisher Scientific.

2.10. Toxigenic Potential of *A. flavus* and *A. parasiticus* Strains for Aflatoxins

The production of aflatoxins was tested using the method of Lin (1976) [33]. The *A. flavus* strains isolated from goat feed (130) and from forage (70) and *A. parasiticus* (01) from feed were maintained in a tube containing PDA at 25ºC for 5 days. A fragment of each colony was then inoculated into a place containing coconut agar medium and incubated at 25ºC for 10 days. After growth, the whole content of each plate was removed, and chloroform was added, 30 mL for each 10 g of culture. After shaking, the sample was filtered through filter paper with diatomaceous earth and sodium sulfate. The filtrated was evaporated, and the extract was suspended in chloroform and then submitted to thin layer chromatography (TLC).

2.11. Aflatoxin Analyses

The aflatoxins identification and quantification were performed by TLC using an aliquot (40 µL) of each sample, which was spotted on silica gel-G thin layer plate (Merck, Germany) and then developed with chloroform: acetone 9:1 (v/v) as a solvent system. The concentration of aflatoxins was determined by photodensitometry (Shimadzu, CS 9000) comparing the area and density of the spot samples with aflatoxins B_1, B_2, G_1, and G_2 standards (Sigma Aldrich, USA) [34]. The detection e quantification limits were established how the lowest fluorescence detectable signal and the lowest concentration measured, respectively.

2.12. Toxigenic Potential of *A. flavus* Strains for CPA

The methodology of GONÇALEZ (2013) [35] was employed to evaluate the CPA production by *A. flavus* strains. The *A. flavus* strains isolated from goat feed (130) and from forage (70) were maintained in a tube containing PDA at 25°C for 7 days. Spore solutions (1.0 mL) of each sample were inoculated into 25 mL of Czapec-Dox broth (Difco) and incubated for 12 days at 25°C. The cultures were filtrated, and the CPA was extracted twice with 25 mL of chloroform. The chloroform was evaporated and the extract was diluted in 1.0 mL of methanol HPLC grade and then submitted to high performance liquid chromatography (HPLC).

2.13. Statistical Analysis

The results were assessed by analysis of variance (ANOVA) and Tukey's multiple range tests with a significance level $P < 0.05$ to determine differences between feed and forage using the variables: fungi isolation, *A. flavus* and *A. parasiticus* identification, mycotoxigenic ability, aw aflatoxins and CPA contamination in feed and their residues in milk.

3. Results and Discussion

Management of animal nutrition has great importance on the milk quality and its by-products. Caring with feeds should extend not only to the nutritional value, but also to their quality, because some compounds, such as cottonseed, cottonseed meal and corn are widely used in animal diet in lactation. They usually are ingredients very susceptible to contamination by mycotoxins and toxigenic fungi. Presence of mycotoxins in food for animal consumption was reported in different regions of Brazil [19, 20, 36]. Of the 128 feed samples analyzed, 2.34% were contaminated with aflatoxins. Aflatoxins were detected in 1 feed sample from P2 property, and in 2 samples from P9 property. Property P2 presented the following aflatoxin concentrations: 36.93 μg/kg AFB_1, 2.46 μg/kg AFB_2, 46.06 μg/kg AFG_1, and 1.65 μg/kg AFG_2, and the sum of the four aflatoxins (87.1 μg/kg) exceeded the maximum limit permitted by Brazilian law for feed ingredients (50 μg/kg – total sum of AFB_1, AFB_2, AFG_1, AFG_2) [37]. Feed samples from P9 property presented the following aflatoxin concentrations: 3.91 μg/kg AFB_1, 1.18 μg/kg AFB_2, 3.58 μg/kg AFG_1, 0.93 μg/kg AFG_2, and 3.65 μg/kg AFB_1; 0.35 μg/kg AFB_2, 1.14 μg/kg AFG_1 and 0.61 μg/kg AFG_2 (Table 1). Batatinha (2008) [38] reported the absence of aflatoxin in 80 feed samples for goats in Brazil.

Surveys regarding the presence of CPA in different foods, including milk, are scarce [23]. Of the 128 samples of feed analyzed, 3.12% were contaminated with CPA, being that it detected only in samples from P1 and P3 properties, but its quantification was not possible since the values were below the quantification limit (3.0×10^{-5} μg/kg). In samples from P8 and P9 properties one found 0.743 μg/g and 1.9 μg/kg of CPA, respectively (Table 1). Brazil does not have the maximum limit permitted for CPA in feed.

Aflatoxin M_1 and CPA were not found in the 120 milk samples analyzed. The absence of AFM_1 and CPA evaluated in all milk samples was positively correlated with the results of the analyses of feed and their statistical significance ($P < 0.05$), which exhibited a low rate of occurrence of AFB_1 and CPA in animal nutrition for the same experimental period. Although three feed samples positive for AFB_1 and four positives for CPA have been encountered AFM_1 and CPA were not detected in their respective milk. The absence of AFM_1 may be related to the rate of elimination from 0.18% to 3.0% of ingested AFB_1. Studies on the elimination of CPA are scanty. Dorner (1994) [11] presented a study in sheep, where from each 5 μg/kg body weight, 236 ng of CPA/g was quantified by HPLC on the first day, hence showing a very low rate of the toxin elimination.

Table 1. Concentration of aflatoxins (μg/kg) and CPA (μg/kg) in feed samples for dairy goats collected from familiar properties in the Southwestern São Paulo region

Properties	Total of samples analyzed (positive)	Mycotoxins (μg/Kg)					
		AFB_1	AFB_2	AFG_1	AFG_2	Total AFs	CPA
P1	17 (0)	ND	ND	ND	ND	-	D
P2	3 (1)	36.93	2.46	46.06	1,65	87.1*	ND
P3	8 (0)	ND	ND	ND	ND	-	D
P4	14 (0)	ND	ND	ND	ND	-	ND
P5	9 (0)	ND	ND	ND	ND	-	ND
P6	19 (0)	ND	ND	ND	ND	-	D
P7	16 (0)	ND	ND	ND	ND	-	ND
P8	7 (1)	ND	ND	ND	ND	-	0.743
P9	18 (2)	3.91	1.18	3.58	0.93	9.6	ND
		3.65	0.35	1.14	0.61	5.75	1.9
P10	17 (0)	ND	ND	ND	ND	-	ND

ND: not detected; D: detected; AFB_1: aflatoxin B_1; AFB_2: aflatoxin B_2; AFG_1: aflatoxin G_1; AFG_2: aflatoxin G_2; Total AFs: aflatoxins (B_1+B_2+G_1+G_2); CPA: cyclopiazonic acid; *Concentration beyond the maximum limit permitted by Brazilian law.

Negative results for the presence of AFM_1 in milk and cheese of goat were also reported by other authors [39, 40]. In Brazil, AFM_1 in goat milk was detected by Barrios (1996) [41], and Oliveira (2007) [22], who observed that 45% and 69.4% of the samples contained AFM_1 at levels ranging from 20-200 μg/kg and 0.011 to 0.161 ppb, respectively. A total of 39 goat milk samples manufactured in three Brazilian states, aflatoxin M_1 was detected in 7 (18%) samples in a range of from 20 to 98 ng/L [42]. In Italy, AFM_1 analysis in samples of milk and cheese of goat showed that about 30% of the samples were contaminated by this toxin at levels from 0.004 to 0.037 μg/kg in milk and from 0.019 to 0.16 μg/kg in cheese [43]. A study carried out in Sardinia in 208 goat milk samples and 41 goat cheese samples detected the presence of AFM1 in 36 (17.3%) at a concentration from 5.0 to 40 ng/L of milk and 9.8% at levels between 79.5 and 389 ng/kg of cheese, respectively [15].

Study of aflatoxin M_1 contamination in milk of four dairy species in Yazd, Iran showed levels of the toxin in 15.4% of cow milk,11.5% of sheep milk, and 9.15% of goat milk samples exceeded the legal limit (0.050 mg/kg) recommended by the Institute of Standards and Industrial Research of Iran [44]. Another study in Iran, AFM_1 contamination was 84.3%, 44.6% and 65.3% for cow, goat and sheep raw milks, respectively. Percentages of cow milk, goat milk, sheep milk and cheese samples exceeding the EU limit were 35.9%, 11.1%, 26.9% and 10%, respectively [24]. Samples were collected from 78 farms, 62 of them were goats and 16 sheep farms in Thessaly, Greece. Out of 234 samples analyzed, in 43 (18.4%) samples AFM_1 was detected, while 4 (1.7%) were found above the maximum tolerable level of 50 ng/kg [45].

Table 2. Relative frequency (%) of the genera fungi contamination of animal feed (128 samples) and forage (106 samples) collected in 10 dairy goat farms in the period May, 2010 to September, 2011

Isolated fungi	Relative frequency (%) in animal feed	Relative frequency (%) in forage
Penicillium spp.	84.5	71.7
Mucor spp.	44.9	30.2
Fusarium spp.	35.6	32.0
Aspergillus spp.	34.1	43.4
Cladosporium spp.	33.3	34.9
N.S.F.	31.0	50.0
Rhizopus spp.	21.0	8.50
Trichoderma spp.	20.9	16.0
Colletotrichum spp.	8.5	26.4
Eurotium spp.	1.5	2.30
Paecilomyces spp.	0.8	-

NSF: Non-sporulating fungi.

An important component of efforts to control mycotoxin contamination problems is the study of the morphological, molecular genetics, metabolic and plant pathological diversity of mycotoxigenic fungi [8]. The mycobiota of the animal feed and forage samples are shown in Table 2. Toxigenic fungi belonging to the genera *Aspergillus*, *Penicillium* and *Fusarium* were isolated from most samples (Table 2). Other studies also reported the presence of *Aspergillus* spp., *Penicillium* spp., *Fusarium* spp., and *Eurotium* spp. in animal feed [46, 47]. Among representatives of the genus *Aspergillus*, were isolated from animal feed: *A. flavus* (69%), *A. fumigatus* (2.30%), *A. niger* (1.50%) and *A. parasiticus* (0.80%). The highest incidence of *A. flavus* was shown in animal feed and in their ingredients [46-48]. The *Aspergillus* species isolated from forage were: *A. flavus* (4 9%), *A. fumigatus* (5.70%), and *A. niger* (1.0%). The forage samples showed highest incidence of *Aspergillus* spp. than in feed. Same results were found by Accensi (2004) [49].

The identification of *Aspergillus* Section Flavi has been traditionally based on morphological and biochemical characterization [2]. The morphological and molecular characterization and mycotoxigenic ability were used for *A. flavus* and *A. parasiticus* identification. Morphologically, *A. flavus* has finely roughened conidia mostly produced from heads bearing both metulae and phialides, whereas conidia of *A. parasiticus* are usually conspicuously roughened and most heads bear phialides alone [32]. Based on these morphological characters, it was possible to identify *A. flavus* in 64% of the *Aspergillus* spp. isolated from feed and 37% from the forage. The identification was confirmed by molecular characterization and mycotoxigenic ability of the *A. flavus* and *A. parasiticus* strains. The results of the molecular analysis showed that 97% of the *Aspergillus* spp. isolated from feed was PCR positive for *A. flavus* and *A. parasiticus*. The molecular identification using the gene *aflD* (PCR) was a good maker for differentiating the *A. flavus* and *A. parasiticus* species (fragment of 400pb) from the other *Aspergillus* spp., but it was not possible to differentiate them from each other (Figure 2). Of 131 strains isolated from feed and confirmed by PCR, 65.60% were *A. flavus* and 0.76% was *A. parasitcus*, using the gene *aflR* (PCR-RFLP). Among the *Aspergillus* spp. isolated from forage (98 strains) 71.40% were positive by PCR analysis and 51.40% of them were identified with *A. flavus*, however *A. parasiticus* was not isolated. The results obtained in this work are in agreement with Somashekar (2004) [3], who showed that PCR-RFLP patterns obtained with *Hinc*II can be used to distinguish the two species. *A. flavus* cleaved into 3 fragments of 385, 250, 161bp whereas *A. parasiticus*, having one restriction site for the *Hinc*II, produced 2 fragments of 546 and 250bp (Figure 3). The *aflD* and *aflR* genes were not able to identify the ability of the strains to produce aflatoxins [2, 3].

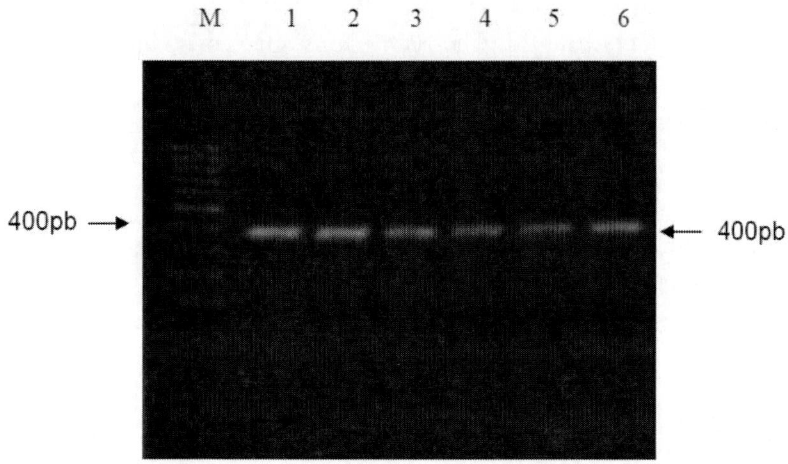

Figure 2. Gel electrophoretic analysis of PCR products. M: molecular 100 pb DNA Ladder (Fermentas); 1-6: *Aspergillus* spp.

In addition, the genomic structural genes involved in aflatoxins biosynthesis do not guarantee the production of aflatoxins by all isolates of *A. flavus* and *A. parasiticus* [4], so their mycotoxigenic ability was investigated. The 131 *Aspergillus* spp. strains confirmed by PCR from feed and 70 strains from the forage were evaluated as having ability to produce aflatoxins and CPA (Table 3). *A. flavus* is able to produce aflatoxins and CPA, and *A. parasiticus* can produce only aflatoxins. The results showed that of the *Aspergillus* spp. from feed 39% produced aflatoxins B_1 and B_2, 17% produced CPA, 18% produced both toxins, and 42% had no toxigenic ability. Only 2.0% of the strains produced aflatoxins B_1, B_2, G_1, and G_2, but no CPA, and, based on all characteristics analyzed, they were identified with *A. parasiticus*. Almost all *A. parasiticus* isolated produce both aflatoxins B and G, but not CPA [50]. The strains from forage were producers of aflatoxins B_1 and B_2 (37%), CPA (14%), both mycotoxins (14%), and 9% did not show toxigenic ability. The concentrations of aflatoxins B_1 and B_2, and CPA were showed in the Table 3. The results showed the great variability in the mycotoxins producing potential by *A. flavus* in the culture conditions. Our results agree with the literature that also found difference in the mycotoxigenic ability of the *A. flavus* [1, 47]. Vaamonde (2003) [1] proposed five chemotypes for *A. flavus* species, based on the mycotoxins produced by them: chemotype I for aflatoxin B and CPA producers; II for AFB, AFG and CPA producers; III for AFB producers; IV for CPA producers and V for non-mycotoxin producers. The *A. flavus* strains isolated from feed and forage belong to the following chemotypes, respectively: type I (18% and 14%); type III (40% and 38%); type IV (17% and 14%) and type V (25% and 35%) (Figures 4 and 5). The chemotype II was not isolated. Astoreca (2011) [47], also isolated *A. flavus* belonging to chemotypes I, III, IV and V, but no type II from poultry feed. By comparing the chemotypes isolated from feed and forage, there was no statically significant difference between them (P < 0.05).

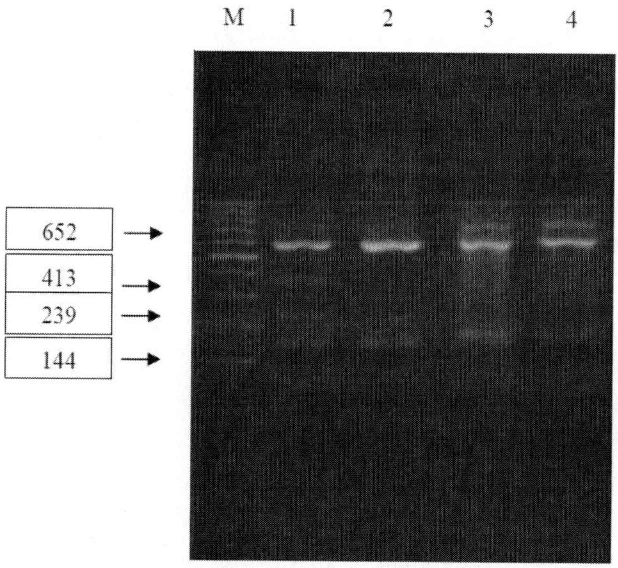

Figure 3. Gel electrophoretic analysis of PCR-RFLP products. M: molecular 100 pb DNA Ladder (Fermentas); 1: *A. parasiticus*; 2-4 *A. flavus*.

Table 3. Aflatoxins B_1 and B_2 and CPA production in culture medium by *Aspergillus* spp. isolated from feed (131 strains) and forage (70 strains) collected in 10 dairy goat farms in the period May, 2010 to September, 2011

	Feed		Forage	
Mycotoxin	**Isolates producer**	**Range**	**Isolates producer**	**Range**
AFB_1	51	9.12 -6,425.43 µg/kg	26	8.35 - 2,944.74 µg/kg
AFB_2	16	7.56 - 87.187 µg/kg	05	3.57 - 49.24 µg/kg
CPA	22	2.12 - 179.06 µg/mL	10	4.10 - 123.92 µg/mL

AFB_1: aflatoxin B_1; AFB_2: aflatoxin B_2; CPA: cyclopiazonic acid.

The average Aw values of the samples ranged from 0.51 to 0.75 for feed and from 0.46 to 1.0 for forage. The minimum Aw values required for growing *A. flavus* and *A. parasiticus* as well as for producing aflatoxins are 0.80 and 0.83, respectively [32] and for CPA production is 0.94. The Aw did not influence the *A. flavus* and *A. parasiticus* isolation from feed and also forage ($P < 0.05$) since these species were isolated in all samples analyses. Therefore, the feed did not have water enough for the fungal activity, so the low level of aflatoxin contamination and CPA in feed samples for feeding goats in lactation may be related to sample water activity. On the other hand, the forage is not a good substrate for mycotoxins production, because it has high fiber concentration and low nutrients to the fungi. Besides that, all properties stored their feeds for a period not exceeding 15 days and in protected locals, free of moisture and rodents, factors that also contributed to the low infection rate of both contaminants. According to Smith (1985) [51], feed with good microbiological quality should present a maximum of 105 colony-forming units/g (CFU/g). The results showed that only 17% of the feed and 2.8% of forage samples had more than 105 CFU/g, therefore most of goat feeding had good quality in all farms studied.

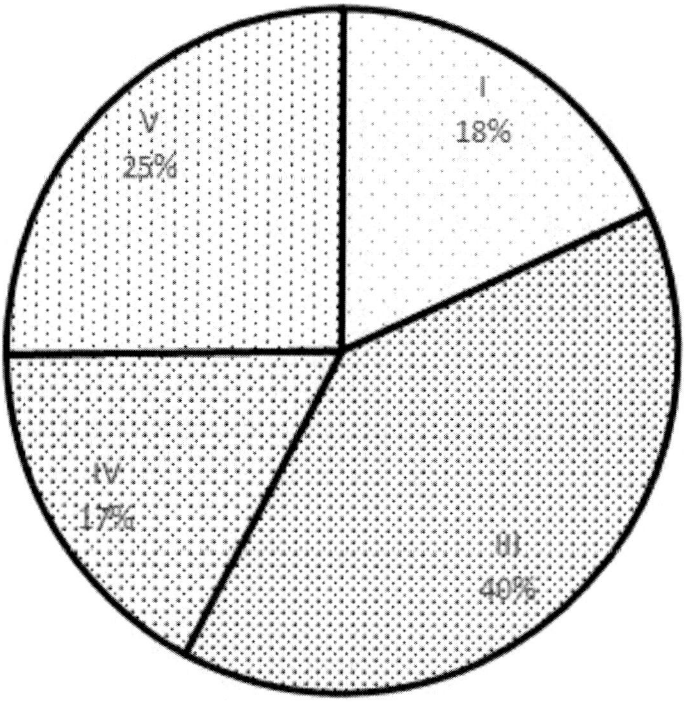

Figure 4. Chemotypes of *A. flavus* isolated from goat feed.

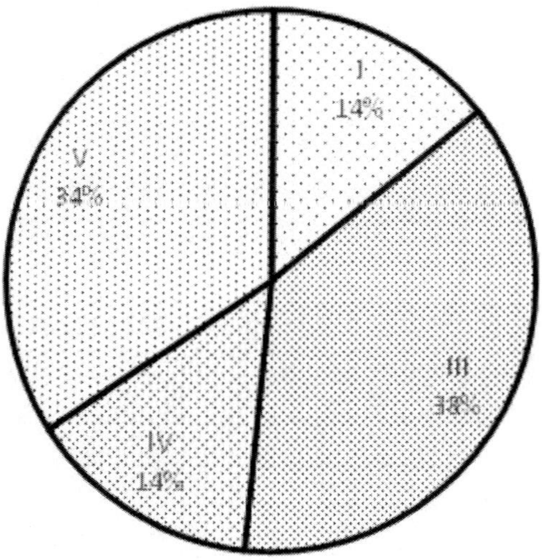

Figure 5. Chemotypes of *A. flavus* isolated from goat forage.

Conclusion

The results obtained in this study indicate that, to distinguish the *A. flavus* and *A. parasiticus* species, it is necessary more than one technique of identification. On despite of low contamination of feed by aflatoxins and CPA and, consequently, their absence in milk, the presence of toxigenic *A. flavus* and *A. parasiticus* indicates a risk of contamination in dairy goat feed if they are exposed to environmental conditions that allow fungal growth. It is very important periodically monitoring the level of mycotoxins in the feed for dairy goats to prevent the milk contamination, once this food is consumed by human, especially children.

References

[1] Vaamonde, G., Patriarca, A., Pinto, V. F., Comerio, R., and Degrossi, C. (2003). Variability of aflatoxin and cyclopiazonic acid production by *Aspergillus* section flavi from different substrates in Argentina. *International Journal of Food Microbiology* 88: 79-84 accessed in 02/15/2008 doi:10.1016/S0168-1605(03)00101-6.

[2] Rodrigues, P., Venâncio, A., Kozakiewicz, Z. and Lima, N. (2009). A polyphasic approach to the identification of aflatoxigenic and non-aflatoxigenic strains of *Aspergillus* section Flavi isolated from portuguese almonds. *International Journal of Food Microbiology* 129: 187-193 accessed in 01/27/2010 doi:10.1016/j.ijfoodmicro. 2008.11.023.

[3] Somashekar, D., Rati, E. R. and Chandrashekar, A. (2004). PCR-restriction fragment length analysis of aflR gene for differentiation and detection of *Aspergillus flavus* and *Aspergillus parasiticus* in maize. *International Journal of Food Microbiology* 93: 101-107.

[4] Levin, E. R. (2012). PCR detection of aflatoxin production fungi and its limitations. *International Journal of Food Microbiology* 156: 1-6.

[5] Gqaleni, N., Smith, J. E., Lacey, J. and Gettinby, G. (1997). Effects of temperature, water activity, and incubation time on production of aflatoxins and cyclopiazonic acid by an isolate of Aspergillus flavus in surface agar culture. *Applied and Environmental Microbiology* 63: 1048-1053.

[6] Smith, T. K. and Korosteleva, S. N. (2012). The significance of feedborne mycotoxins in ruminant nutrition. In: Gonçalez, E., Felicio, J. D., Aquino, S. (editors). *Mycotoxicoses in animals economically important*. New York: Nova Science Publishers, Inc., 2012, pp.35-66.

[7] Osweiler, G. D. Equine mycotoxicosis. In: Gonçalez, E., Felicio, J. D., Aquino, S. (editors). *Mycotoxicoses in animals economically important*. New York: Nova Science Publishers, Inc., 2012, pp.1-34.

[8] Moretti, A. (2013). Molecular biodiversity of mycotoxigenic fungi that threaten food safety. *International Journal Food Microbiology* 167: 57-66.

[9] Oliveira, C. A. F. and Corrêa, B. (2012). Interactive effects between mycotoxins in livestocks. In: Gonçalez, E., Felicio, J. D., Aquino, S. (editors). *Mycotoxicoses in animals economically important*. New York: Nova Science Publishers, Inc., 2012, pp. 114-125.

[10] Ismail, M. A. (2000). Deterioration and spoilage of peanuts and desiccated coconuts from two sub-saharan tropical East African countries due to the associated mycobiota and their degradative enzymes. *Mycopathologia* 150: 67-80 accessed in 04/22/2007 doi.org/10.1023/A:1010863507652.

[11] Dorner, J. W., Cole, R. J. C., Erlington, D. J., Suksupath, S. and McDowell, G. H. (1994). Cyclopiazonic Acid Residues in Milk and Eggs. *Journal of Agricultural and Food Chemistry* 26: 1516-1518. accessed in 09/07/2010 doi.org/10.1021/jf00043a023.

[12] Rosmaninho, J. F., Oliveira, C. A. F., Dos Reis, T. A. and Corrêa, B. (2006). Aflatoxina M1 e Ácido Ciclopiazônico em Leites de Consumo Comercializados no Município de São Paulo, SP, Brasil. *Brazilian Journal of Food Technology* 3: 55-59. [Aflatoxin M1 and cyclopiazonic acid in milk form São Paulo city, SP, Brazil. *Brazilian Journal of Food Technology* 3: 55-59].

[13] International Agency for Research on Cancer – IARC. (2002). Some traditional herbal medicines, some mycotoxins, naphthalene and styrene. Summary of data reported and evaluation. In: *IARC monographs on the evaluation of the carcinogenic risk to humans*. Lyon: IARC Scientific Publications, 82, pp. 171-175.

[14] Hussein, H. S. and Brasel, J. M. (2001). Toxicity, metabolism, and impact of mycotoxins on humans and animals. *Toxicology* 167,101–134.

[15] Virdis, S., Corgilo, G., Scarano, G., Pilo, A. L. and De Santis, E. P. L. (2008). Occurrence of aflatoxin M1 in tank bulk goat milk and ripened goat cheese. *Food Control* 19: 44-49.

[16] Iqbal, S. Z., Jinap, S., Pirouz, A. A. and Faizal, A. R. A. (2015). Aflatoxin M1 in milk and dairy products, occurrence and recent challenges: A review. *Trends in Food Science & Technology* 46: 110-119.

[17] Tsakiris, I. N., Tzatzarakis, M. N., Alegakis, A. K., Vlachou, M. I., Renieri, E. A. and Tsatsakis, A. M. (2013). Risk assessment scenarios of children's exposure to aflatoxin M1 residues in different milk types from the Greek market. *Food and Chemical Toxicology* 56: 261-265.

[18] Galvano, F., Galofaro, V. and Galvano, G. (1996). Occurrence and stability of aflatoxin M1 in milk and milk products: a worldwide review. *Journal of Food Protection* 59: 1079-1090.

[19] Gonçalez, E., Pinto, M. M., Manginelli, S. and Felicio, J. D. (2004). Intoxicação de vacas leiteiras por farelo de algodão naturalmente contaminado com aflatoxinas. *Ciência Rural* 3: 171-174. [Dairy cows poisoned with cottonseed meal naturally contaminated with aflatoxins. *Ciência Rural* 3: 171-174].

[20] Oliveira, C. A. F., Sebastião, L. S., Fagundes, H., Rosin, R. E. and Fernandes, A. M. (2008). Aflatoxins and cyclopiazonic acid in feed and milk from dairy farms in São Paulo, Brazil *Food Additives and Contaminants: Part B* 1: 147-152.

[21] Motawee, M. M., Bauer, J. and McMahon, D. J. (2009). Survey of Aflatoxin M1 in Cow, Goat, Buffalo and Camel Milks in Ismailia-Egypt. *Bulletin of Environmental Contamination and Toxicology* 83: 766-769.

[22] Oliveira, C. A. F. and Ferraz, J. C. O. (2007). Occurrence of aflatoxina M1 in pasteurized, UHT milk and milk powder from goat origin. *Food Control* 18: 375-378.

[23] Flores-Flores, M. E., Lizarraga, E., López De Cerain, A. and González-Penas, E. (2015). Presence of mycotoxins in animal milk: a review. *Food Control* 53: 163–176.

[24] Bahramia, R., Shahbazia, Y. and Nikousefat, Z. (2016). Aflatoxin M1 in milk and traditional dairy products from west part of Iran: Occurrence and seasonal variation with an emphasis on risk assessment of human exposure. *Food Control* 62: 250–256.

[25] García, V., Rovira, S., Boutoial, K. and López, M. B. (2014). Improvements in goat milk quality: A review. *Small Ruminant Research* 121: 51–57.

[26] Park, Y. W., Júarez, M., Ramos, M. and Haenlein, G. F. W. (2007). Physicochemical characteristics of goat and sheep milk. *Small Ruminant Research* 68: 88–113.

[27] Jandal, J. M. (1996) Comparative aspects of goat and sheep milk. *Small Ruminant Research* 22: 177–185.

[28] Chan, D., Macdonald, S. J., Boughtflower, V. and Brereton, P. (2004). Simultaneous determination of aflatoxins and ochratoxin A in food using a fully automated immunoaffinity column clean-up and liquid chromatography-fluorescence detection. *Journal of Chromatography A.* 1059: 13-16.

[29] Tarin, A., Rosell, M. G. and Guardino, X. (2004). Use of high-performance liquid chromatography to assess airbone mycotoxins aflatoxins and ochratoxin A. *Journal of Chromatography: Part A* 1047, 235-240.

[30] Dragacci, S., Grosso, F. and Gilbert, J. (2001). Immunoaffinity column cleanup with liquid chromatography for determination of aflatoxin M1 in liquid milk: collaborative study. *Journal of AOAC Internatinal* 84: 437-443.

[31] Prasongsidh, B. C., Kailasapathy, K., Skurray, G. R. and Bryden, W. L. (1998). Analysis of cyclopiazonic acid in milk by capillary electrophoresis. *Food Chemistry* 61: 515-519.

[32] Pitt, J. I. and Hocking, A. D. (1997). *Fungi and food spoilage*. New York: Springer, p. 506.

[33] Lin, M. L., and Dianese, L. C. A. (1976). Coconut agar medium for rapid detection of aflatoxin production by *Aspergillus* spp. *Phytopathology* 66: 1466-1469.

[34] Gonçalez, E., Felicio, J. D. and Pinto, M. M. (2001). Biflavonoids inhibit the production of aflatoxin by Aspergillus flavus. *Brazilian Journal of Medical and Biological Research* 34: 1453-1456.

[35] Gonçalez, E. Silva, J. L., Reis, T. A., Nakai, V. K., Felicio, J. D. and Corrêa, B. (2013). Produção de aflatoxinas e ácido ciclopiazônico por cepas de *Aspergillus flavus* isoladas de amendoim. *Arquivos do Instituto Biológico* 80: 312-317. [Production of aflatoxin and cyclopiazonic acid by *Aspergillus flavus* strains isolate from peanuts *Arquivos do Instituto Biológico* 80: 312-317].

[36] Sassahara, M., Yanaba, E. K. and Netto, D. P. (2003). Ocorrência de aflatoxinas e zearalenona em alimentos destinados ao gado leiteiro da região norte do Estado do Paraná. *Semina: Ciências Agrárias* 24: 63-72. [Occurrence of aflatoxins and zearalenone in feedstuffs destined to dairy cattle consumption in Northern Parana, Brazil. *Semina: Ciências Agrárias* 24: 63-72].

[37] Brasil. (1988). *Ministério da Agricultura, Pecuária e Abastecimento*. Portaria MA/SNAD/SFA No. 07, de 09/11/1988, Diário Oficial da União de 09 de novembro de 1988, Seção I, pp. 21.968.

[38] Batatinha, M. J. M., Botura, M. B., Simas, M. M. S., Mallmann, C. A., Bittencourt, T. C. B. S. C. and Sabino, M. (2008). Otimização de método analítico para determinação de aflatoxinas em rações de caprinos e ocorrência no estado da Bahia. *Arquivo Brasileiro de Medicina Veterinária e Zootecnia* 60: 503-505. [Optimization of analytical method for aflatoxins determination in goat's feedstuffs and their occurrence in the state of Bahia. *Arquivo Brasileiro de Medicina Veterinária e Zootecnia* 60: 503-505].

[39] Botura, M. B. (2005). Otimização de métodos analíticos para determinação de aflatoxinas em rações e leite de cabra, e sua ocorrência no Estado da Bahia. Escola de Medicina Veterinária da Universidade Federal da Bahia, Bahia, Brazil. [Optimization of analytical method for aflatoxins determination in goat's feedstuffs and their occurrence in the state of Bahia, master's thesis].

[40] Tsiplakou, E., Anagnostopoulos C., Liapis, K., Haroutounian, S. A. and Zervas, G. (2014). Determination of mycotoxins in feedstuffs and ruminant's milk using an easy and simple LC–MS/MS multiresidue method. *Talanta* 130: 8-19.

[41] Barrios, M. J., Gualda, M. J., Cabanas, J. M., Medina, L. M. and Jordano, R. (1996). Occurrence of aflatoxin M1 in cheeses from de south of Spain. *Journal of Food Protection* 59: 898-900.

[42] Iha, M. H., Abe, L. T., Okada, M. M. and Fávaro, R. M. D. (2007). Occurrence of aflatoxins M1 and M2 in goat milk marketed commercialized in the region of Ribeirão Preto-SP, Brazil. *Revista do Instituto Adolfo Lutz* 66: 57-62.

[43] Finoli, C. and Vecchio, A. (1997). Aflatoxin M1 in goat dairy products. *Microbiologie-Aliments-Nutrition*: 15, 47-52.

[44] Fallah, A. A., Fazlollahi, R. and Emami, A. (2016). Seasonal study of aflatoxin M1 contamination in milk of four dairy species in Yazd, Iran. *Food Control* 68: 77-82.

[45] Malissiovaa, E., Tsakalofa, A., Arvanitoyannisd, I. S., Katsafliakaa, A., Katsioulisa, A., Tserkezoua, P., Koureasa, M., Govarise, A. and Hadjichristodouloua, C. (2013). Monitoring Aflatoxin M1 levels in ewe's and goat's milk in Thessaly, Greece: potential risk factors under organic and conventional production schemes. *Food Control* 34: 241-248.

[46] Pereyra, C. M., Cavaglieri, L. R., Chiacchiera, S. M. and Dalcero, A. M. (2010). Fungi and mycotoxins in feed Intended for Sows at different reproductive stages in Argentina. *Veterinary Medicine International* ID 569108 accessed in 08/26/2011 doi:10.4061/ 2010/569108.

[47] Astoreca, A. L., Dalcero, A. M., Pinto, V. F. and Vaamonde, G. (2011). A survey on distribution and toxigenicity of *Aspergillus* section Flavi in poultry feeds. *International Journal of Food Microbiology* 146: 38-43.

[48] Pereyra, C. M., Cavaglieri, L. R., Chiacchiera, S. M., Dalcero, A. M. (2011). Mycobiota and mycotoxins contamination in raw materials and finished feed intended for fattening pig's production in eastern Argentina. *Veterinary Research Communication* 35: 367-379 accessed 06/06/2012 doi: 10.1007/s11259-011-9483-9.

[49] Accensi, F., Abarca, M. L. and Cabañes, F. J. (2004). Occurrence of *Aspergillus* species in mixed feeds and component raw materials and their ability to produce ochratoxin A. *Food Microbiology* 21: 623-627.

[50] Horn, B. W. and Dorner, J. W. (1999). Regional differences in production of aflatoxin B1 and cyclopiazonic acid by soil isolates of *Aspergillus flavus* along a transect within the United States. *Applied and Environmental Microbiology* 65: 1444-1449.

[51] Smith, J. and Moss, M. (1985). *Mycotoxins, formation, analysis and significance*. New York: John Wiley and Sons, p.223.

INDEX